LEONARDO'S VISION

A guide to collective thinking and action

Valerie A. Brown

A C.I.P. record for this book is available from the Library of Congress.

ISBN 978-90-8790-134-9 (paperback)
ISBN 978-90-8790-135-6 (hardback)

Published by: Sense Publishers,
P.O. Box 21858, 3001 AW Rotterdam, The Netherlands

http://www.sensepublishers.com

Judith Wright: 'Eve to Her Daughters' and 'Fission and Fusion' from A Human Pattern: Selected

Poems (ETT Imprint, Sydney 1996).

Illustrations © Photo SCALA, Florence 2006

Cover design and chapter headings based on ImageAnthroposX.jpg - Wikimedia

Book design: Kim Tatnell, Big Island Graphics

Printed on acid-free paper

To Robin Wallace Brown, with love

Contents

Foreword

Valerie Brown, as an inspirational educator and trailblazing social change agent for decades, has written a book on social learning for transforming society. This requires society covering all knowledge bases so to speak. All forms of knowledge and cultures have to be conscientiously covered in society for sustainability to have a chance. Valerie's book has good examples of the processes by which this can occur based on extended practical experience. From now on in society, she is arguing, each generation will need to learn to become more empathetic and competent with each of the knowledge cultures. Eventually society as a whole will have to have some kinds of formal processes and dedicated places to enable this to occur. Valerie's book is about transforming society from within by transforming the ways we think and act in the human, and the more than human world. It is an ideal. It is full of hope and optimism.

It is Leonardo's gift or his inspiration that is also at the heart of this book. Valerie has shared how she was (and is) so much inspired by Leonardo da Vinci and his successors that she decided to write the book. This personalises the book for the reader and enables Valerie to incorporate her own experience into the book. She has devoted much of her life to breaking down barriers to learning and living by building up dialogue skills between people in order for them to learn to share their knowledge and understanding. It has been Valerie's way of empowering people, increasing their capacity to learn, be heard and act responsible as individuals and as members of a social group, local community, and the wider society. Once learning is empowered in people (often after a dis-empowering formal education) they become more ready, willing and able to work things out for themselves and to take appropriate action.

The long-term goal in the book is for a just, secure and sustainable future. It is not a realistic goal in the current socio-political context but a felt necessity and a necessary statement of hope for many of us. Could global climate change be the catalyst for setting out on Valerie's way ahead? Then Valerie's emphasis on social learning for transforming society could be off and running. Here is a handbook for implementing transformative learning throughout the whole of society by acknowledging that there is a different way ahead. It seems that Leonardo's vision is about learning to share the space we have claimed for ourselves as individuals, members of knowledge cultures, national governments, etc. Further, the more of humanity there are, the less space there is for others including our fellow life travelers on Earth. This is what a just, secure and sustainable future is all about too: how do we share our planet? It requires the socially transformative kinds of learning which are offered in Leonardo's vision.

John A. Harris BSc, PhD, Ecologist.
Coordinator, Master of Community Environmental Studies
University of Canberra,
Australia, 2006

Prologue: in the steps of Leonardo

"Leonardo da Vinci, 1452-1519, a true Renaissance man, whose habit of rigorous enquiry, observation, and experiment, grounded in a philosophic system, led him to conceive of the universe as an organized cosmos corresponding to a work of art."

- Kenneth Clark, 1935. 'Introduction' Leonardo da Vinci London Penguin.

This book has been written in the belief that learnintg to act collectively with a shared understanding of the whole is the major challenge of our time. Our society and our lives have become structured in ways that actively prevent us working together towards common goals. Institutionalised barriers block collaborative action on crucial matters of economic and environmental management, health and future community well-being. A strong antidote is needed in order to access the immense capabilities hidden behind the socially-embedded divisions and oppositions that limit much of what we do.

As individuals, competing personal, professional and organisational loyalties can fragment our everyday lives. Concentrating on specialised skills restricts the development of the full set of our abilities. Communities struggle to maintain a shared identity in the face of constant social change. Organisations have licence to bypass the public good and work only for their private ends. Governments seek to reconcile competing economic, social and environmental interests. Globally, a humane future for Earth requires a concerted response to the planet's disintegrating life support systems. We need to ask whether it is possible to re-order these powerful divisive systems into a different pattern, one based on collective decisions and collaborative action.

A positive answer lies right under our noses, and is well within our existing capabilities. Successful resolutions of the complex issues surrounding the future of both people and planet are already available. Some of them make up the content of this book. But the difficulties should not be underestimated. Collective thought and action requires a two-fold transformation in our current ways of thinking and acting.

First, as individuals we need to bring together our mutual learning from the many facets of a changing world, while maintaining our personal integrity. Second, as a society we need the tools and the courage to overcome the social divisions that stand in the way of a humane and sustainable future, while building on our learning from the past.

To take either of these actions requires significant change in how we acquire our knowledge and make our decisions. Our approved mode of thinking has been to reduce issues into their component parts and focus on one aspect at a time. Our social institutions then ensure that our lives are lived within the restrictive limits set by that thinking. The task before us is not to reject the previous thinking out of hand, but to re-position it within a wider perspective that connects the parts and throws light on the whole.

Such ground-breaking shifts have happened several times in history, although never over a canvas as wide as the future of the planet. One such shift, the innovative era of the 15th century Renaissance, grew out of a simultaneous openness to learning in all branches of knowledge. This same human capacity to think and act across boundaries is available to address the complexity of the issues that face the 21st century, even though under very different circumstances.

One of the pathfinders of the Renaissance, and an icon of our own time, Leonardo da Vinci offers a bridge between the two eras. He was a person whose world was not divided. His paintings carry strong political messages as well as being superbly beautiful (think of The last supper). His inventions helped open up whole new worlds of human experience (the aeroplane, the diving suit, irrigation canals and the bicycle). His journals foretold the discovery of unifying threads of thought, such as chaos theory and the double helix of DNA. Even in his scribbling he bridges the boundaries we have erected between interpretations of the physical and social worlds, between accuracy and imagination (as in the figure below). He continually faced disapproval and even rejection by the institutional forms of his day, at the same time as being the object of adulation in his own community.

To 21st century minds trained to keep their thinking in different compartments, Leonardo's ability to combine diverse capacities in understanding the whole can seem beyond their grasp. Yet the events reported in this book confirm that each human being is born with the same potential for integrative thinking as Leonardo and his contemporaries. Each community holds the same inherent capacity to bring all its capacities to bear on moving towards a better future.

An increasing number of people are making their contribution towards a more united world. This may be through being a Renaissance thinker and writer; in a formal role as negotiator, strategic planner, knowledge broker, advocate, or team-builder; or as a personal choice. From whatever starting point, finding the connecting patterns in today's world is no straightforward task. My own experience of the difficulties began after I completed one of the earliest ecology courses in the 1950s. I had learnt to regard the world as a dynamic living system. In my first job, I was one of a team conserving the natural systems of a sub-tropical rainforest. I soon found that any success lay in combining good science and strong skills in the socio-political arena. After realising that both fields are part of the one dynamic system, I also learnt the difficulty of finding a meeting-place between the two.

Over the next ten years I applied this experience to the public health issues of a small city, at a time when the impacts of AIDS, recreational drugs, and alienation of youth were making their mark. The barriers to achieving constructive solutions proved to be deeply embedded in Western social systems. So I attempted to discover what it would take for whole-of-community involvement in constructive social change. In a wide-ranging action research study, the Local Sustainability Project, my fellow investigators were the initiators of successful integrative programs in highly dispersed fields of action. There I was able to observe the ways of restoring the connections between individuals and their society, community and government, experts and politicians, social and economic resources, the sciences and the arts, and local and global scales that could make up a portfolio of collective practice.

We may not be able to go back to the Renaissance, but this century has strengths of its own. Renaissance thinkers of our own time, from Charles Darwin through to Mahatma Gandhi and Pablo Picasso, and the many less well-known synthesists, have helped break through the partitions imposed by society. The great information highway of the Internet offers transparent pathways ripe for mutual learning. We have the challenge of the times, the pathfinders, the communication technology, and the human intellectual capacity to set the present generation free to find new ways to resolve the immense problems they have inherited.

Figure. Leonardo da Vinci *Drawings of figures*. Florence,

Uffizi © Photo SCALA

In describing all this, there is need for a transparent language that everyone in the community can understand. Throughout the book, specialised terms and ideas are either explained or translated into standard English. Individuals who have introduced new ideas are clearly identified. Acronyms are used rarely. Ideas are presented through text, detailed boxes, examples, stories, paintings, poetry, diagrams, and internet sites. The many ideas for advancing collective thinking and action that can only be briefly referred to here are listed in the bibliography for the reader to pursue further.

The book that follows brings together a portfolio of ideas, strategies and tools that support collective decision-making among individuals, communities, professions and organisations. Since the tools must challenge the very basis of the way things are done in a fragmented society, their use often generates conflict and denial. This in turn calls for personal and professional courage in struggling with the new ideas. I found that changing one's own way of thinking was accompanied by a loss of confidence, uncertainty, and a frantic effort to position oneself in the new knowledge landscape. The reward is the inspiration of working with the many other practitioners of synthesis that have managed to initiate collective ways forward.

As far as we know, Leonardo left no direct descendants. But all those who work towards a collective understanding followed by collaborative action carry his vision forward.

Valerie A. Brown, Canberra 2006

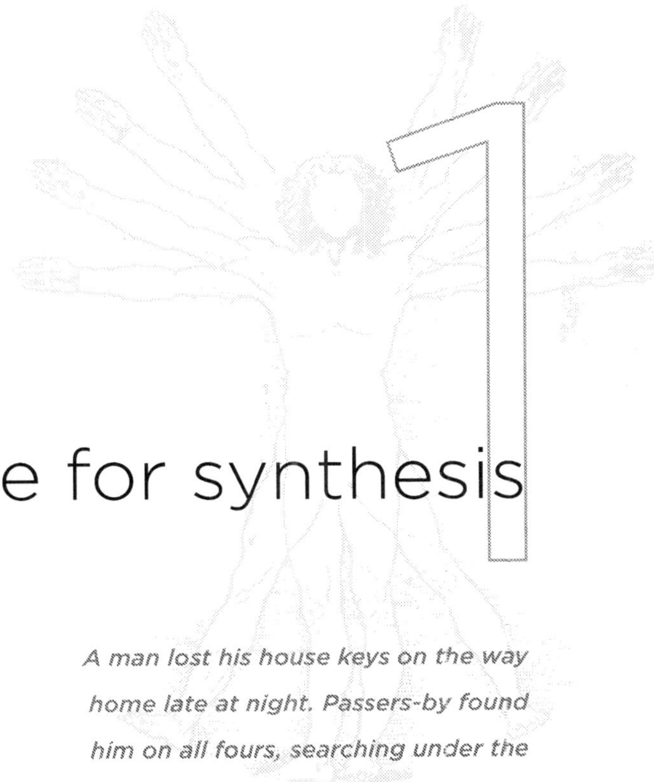

1 The case for synthesis

A man lost his house keys on the way home late at night. Passers-by found him on all fours, searching under the lone street light. After helping him look for some time, someone said "Are you sure you lost them here?" "Oh no," said the man, "I lost them over there. But it's not worth looking there. There's not enough light."

In the dominant form of Western culture, the light shines brightly on analysis, competition, individualism and hierarchical power systems. Tools for synthesis-based thinking, whole-of-community engagement, collaborative inquiry, and integrative management lie in the shadows, outside everyday practice, But for those seeking to guide social change towards a more humane and sustainable future, it is there in those shadows that they will find their way.

Where have we been?

We live in a globalised era, but a divided world. Partitioned relationships, territories and governments have become the standard conditions of our everyday life. Strongly defended boundaries protect authoritarian power, impair the integrity of our social systems and impede the sustainable management of resources. We are so accustomed to living with this all-pervading divisiveness that we no longer notice that it is undermining our capacity to live socially unified lives in an environmentally sustainable world.

In such a fragmented world, anyone who accepts the responsibility for building connections is taking part in a magnificent experiment, one that needs creativity and courage. Strong powers are protecting the different compartments into which society has become divided. In *Guns, germs and steel*, Jared Diamond documents how the analytic tools of the scientific Enlightenment have re-shaped the planet over the past four hundred years. The same specialised and detailed mode of inquiry shapes and confines our social institutions. Equally powerful synthesis-based tools are needed to heal the resulting disruption of the world's social and environmental systems.

It seems that we don't have many more years to find those synthesis-based tools and learn to use them effectively. Forecasters such as Tony McMichael in *Human frontiers, environments and disease* in 1999 and Lester Brown in *Vital signs 2005* give us less than 30 years to arrive at the unified action needed to ensure a humane future for Earth. Fortunately, movement in that direction has already begun. Many successful examples of collective thought and action make up the content of this book.

However, those examples only confirm that the task of accepting synthesis as a primary goal is far from straightforward. The dominant Western analytic approach to knowledge places many hurdles in the way of those trying to redress the prevailing fragmentation of issues and ideas. Strict rules of analysis and hierarchical structures have become firmly established as the default option for structuring how we think. Since the era dominated by the scientific thinking that emerged nearly 4 centuries ago, it has become the prevailing rule to approach any complex problem through its constituent parts. This sets the pattern followed by individuals, government, industry and communities alike. The Nobel-prize-winning chemist Peter Medawar called this approach to problems 'the art of the soluble', finding solutions more readily through addressing the part rather than the whole

The move to treat the complex issues of a sustainable humane future as open-ended and interconnected systems, rather than reduce them to single factors and fixed hierarchies, demands a re-alignment of the social structures developed to support the divisions. The task at hand is to establish that re-alignment while maintaining all the advantages of specialised inquiry.

Calls for integrated decisions and actions that support constructive social change continually refer to the need to bring together the key decision-making groups of community, specialists and government. Each of these groups has its own problems, however. Each group is divided from within. Communities are becoming increasingly stratified by age group, income level and ethnicity, and losing their sense of a shared vision for the future. Specialisations are highly skilled in one particular way of approaching a problem, so that their advice is based on only a partial view of the world. Organisations, both business and government, increasingly exert their control over citizens who are supposedly set free to act individually through democratic processes. As we shall find from the stories from the field, the differences between the three sectors are great enough to amount to different constructions of reality. Lasting mutual decision-making requires active promotion of collective thinking within the groups, as well as between them, and the capacity to put that thinking into effect.

This is not the first time that there has been a search for ways to unite the many different parallel worlds. The early Greek philosophers, from Socrates to Plato to Aristotle, debated the connections between the one and the many, the parts and the whole. Socrates sought the whole; Aristotle classified the parts. The anthropologist/biologist/psychologist Gregory Bateson described the world as formed by the patterns that connect, not the walls that divide. In *Steps to an ecology of the mind*, he points out that the current understanding of the physical world predetermines the way each era shapes its own social world.

Half way in time between Socrates and Bateson came the 15th century Renaissance with its openness to synergies among different ways of thinking. The result was the surge in the knowledge generated in science, finance, management, art and literature. The current popular interest in Leonardo and his times over 400 years later reflects the strength of the desire to reform a compartmentalised world.

The conditions for the release of the integrative energy of the Renaissance were very different from the globalised world of today, but no more promising of success. Then, the background was warring private armies, relentless religious persecution and raging bubonic plague. Now, the conditions include weapons of mass destruction, blind faith in a limitless natural environment and the seeds of pandemics such as bird flu and SARS. But the vision of both eras is the same: harnessing the whole power of the human capacity for collective thinking and action, and freedom from the restrictive walls of dominating institutions.

The Jesuit mammalian archaeologist Teilhard de Chardin in his 1955 *The phenomenon of man* argues that the next human evolutionary advance will be towards a collective brain, able to draw all human capacities together as one, while still respecting the individual mind. This prediction is proving less far-fetched as time goes by. The synthesising power of the information revolution has come into being. Formal positions as coordinator, integrator and synthesiser are increasingly added to the existing range of specialised tasks in management, research, and education. Participatory workshops, conferences and consultations are core management and educational tools. Alliancing has become the preferred mode of management in enterprising organisations.

The prospect of contributing to such concerted social change invites serious questions:

1. How would adopting the goal of synthesis alter our thinking, our behaviour and our institutions?
2. Can the changes be absorbed into present ways of constructing the world, or is the outcome social transformation?
3. Could the changes amount to a new renaissance, bringing into being a global village in thought and action?
4. If synthesis becomes central to thinking and action, is that sufficient in itself to guide us to a more sustainable future?

These are big questions. For a positive answer to the last question, we can turn to the conclusions of the 1999 report *Our common journey: the transition to sustainability*, issued by the combined science, health and engineering research institutes of the United States Research Council. The urgency for synthesis noted there is reinforced by the 2005 Millennium Ecosystem Assessment, *Biodiversity and human well-being: a synthesis report*, based on work by 1300 experts from 95 countries. These authoritative reports agree that arresting the spiral of environmental disruption cannot be achieved in piecemeal fashion or by any one set of interests alone. A new type of community-based collective thinking is required.

This book attempts to answer the first three questions by combining the work of current renaissance thinkers and field studies of collective decision-making. Studies conducted during 1996-2003 taught us not to underestimate the task. The fragmentation we need to repair is not an accident. It is central to the current dominant cultural approach to knowledge. Four centuries of developing analytical skills – of concentrating on ever more detailed parts of an issue – has led to the triumph of Western technology. It has also established the current mode of divided decisions responsible for the inability to act on many warnings about the future.

Where are we going?

Retracing our steps to the words of Leonardo da Vinci, celebrated as painter, sculptor, architect, musician, engineer, mathematician, politician, philosopher and scientist:

"Everything comes from everything, and everything is made from everything, and everything can be turned into everything else."

- Leonardo da Vinci, Notebooks

Cryptic in their own time, these words were written four centuries before Niels Bohr discovered the structure of the atom that underlies all matter, and James Watson the double helix of DNA that carries the code for all inheritance. Those discoveries confirm just how accurate Leonardo was, and how ideas of connectedness persist across time. Einstein's understanding of relativity led to his conclusion that it is not possible to create change in society from within the thinking that created the need for change in the first place. We have come to one of the points in history where we need to bring together what we know in new ways and new patterns capable of grasping the whole. We have become supremely successful at pulling things apart. Putting them together seems to be a mental discipline we hope to avoid.

To date, remedial efforts to bring collective thinking into the mainstream involve an expert negotiator here, or a community newsletter there. There are growing numbers of multidisciplinary research projects, interdepartmental committees, and government coordinating programs. Consensus conferences, citizen's juries and multidisciplinary forums are opening up new forms of collective decision-making. While all of these are steps in the right direction, none have proved enough by themselves. As long as the efforts remain piecemeal and short term, they merely act to uphold the existing divisions.

It is not a question of synthesis replacing analysis, but of whether we can reclaim the capacity to connect the two, a capacity last commonplace during the Renaissance. Unfortunately, in today's society the contributing parts of any collective decision are treated as competing interests, rather than as the interactive parts of a single evolving pattern. To alter this in pursuit of a mutual understanding and collective action requires a new set of skills.

Box 1.1
SUCCESSFUL SYNTHESES *

- A winning football team
- A profit-making corporation
- The city of Paris
- A powerful painting
- Acetylated salicylic acid (aspirin)
- White light
- Dynamite
- An orchestra
- A street gang
- Effective executive government
- A creative leap

* Oxford Concise Dictionary: *synthesis: the fusion of separate elements into a new whole*

In a successful synthetic system, such as the examples in box 1.1, interaction between the parts creates synergy - that is, their interaction leads to a more effective result than any of the parts acting alone. An orchestral performance depends on every musician both acting on their own and following the conductor. Cocktails of chemicals give us the pain relief of aspirin, the power of dynamite and a treatment for AIDS, that no one of the ingredients can supply alone. A city is the outcome of a continuing dynamic interaction between its people, place and resources. It is not too much to say that the creative energy from synthesis is a primary driver of the whole of society.

While all syntheses offer fresh potential, their value for positive change rests on their ethical purpose. Synthesis - whether of chemicals, information, ideas, designs or decisions - is a powerful tool. The worth of the synthesis, however, lies in the use to which it is put. Dynamite can be used in mining and in war. A city can be heaven or hell as a place to live. There are three core issues in developing the capacity for collective thinking and action towards a sustainable future: scoping the context, clarifying the principles and designing the tools. Chapters 2 and 3 scope the context, and chapter 4 introduces the reader to potential principles and tools. The tools are then applied in synthetic decision-making for individuals, communities, professions and organisations in the chapters that follow.

Learning from Google

Among the effective tools for synthesis already at hand, the search engine Google is a prime example. At the turn of the millennium, it emerged in cyberspace as the first 'relational' search engine, using connections rather than divisions to bring its information together. Today there is a verb - to google - meaning to access the full range of ideas that can expand understanding of a topic.

Try googling the word 'synthesis' for yourself, right now. The knowledge gained from a synthesis-based system like Google is very different to that of an analytic system such as a dictionary or reference library. In an analytic system the information gained flows sequentially from a given topic. You choose a topic A, look for a definition for it in B, and follow the trail to sub-topic C, then to sub-sub topic D, and possibly even further into detail, until you find the precise application for which you were looking.

A search with synthesis in mind is quite different. The googled themes in figure 1.1 are not a random list of unrelated items. They form an interconnected pattern. That pattern includes science, art, politics, information technology, social customs and environmental systems. Every theme makes a different contribution to the whole. Taken separately, the items lead to 12 different interpretations of synthesis. Taken together, they represent the key dimensions of change in Western culture at this time.

Figure 1.1 Google search engine responses to 'synthesis'

Fields:	Theme:
Art	Collage: new images from apparently unrelated materials
Astronomy	Finding new solar systems
Computing a.	Designing open access systems, relational systems
Computer b.	Convergence of voice and text, and sight and sound
Education	Life-long learning
Environment a.	Ecology: air, soil, water, and energy flows in living systems
Environment b.	GIS: computer-based geographic information systems
History	Tension between the universal and the particular
Literature	Relationships between parts and wholes
Mathematics	Fourier series of sound waves
Multiculturalism	Celebration of diversity
Psychology	Gestalt psychology: the creative leap
Philosophy	Reconciliation of oppositions: thesis, antithesis, synthesis
Science a.	Critical review of the literature on a topic
Science b.	Generating new organic molecules
Social systems	Social ideals of democracy, greens, feminism, anarchy........
Sociobiology	Biological basis for human society, called 'the new synthesis'
Social ecology	Interactions between social and environmental systems

Source: http://www.google.com.au/ [accessed 5.01.07]

The Google exercise demonstrates that synthesis is a way of thinking that permeates all forms of knowledge, as well as an outcome in its own right. Finding the whole from the Google search includes discovering new galaxies and inventing new chemical compounds, as well as different ways of organising society and managing the environment. However, retrieving that information does not guarantee synthesis in thinking will follow.

The almost unlimited technical capacity for synthesising information offered by the new information technology is invaluable as a tool, but almost useless by itself. The human brain can do more than any computer system in assimilating, sorting and going beyond the available evidence. Maximising the human capacity to work towards shared ethical principles, and releasing the synergy liberated by collective thinking, is the real task.

Time for another Renaissance

There is the mechanical means, and major social and scientific justification for moving towards collective thinking. The increasing interest is demonstrated in the growing fascination with everything we can learn from the prototype of the Renaissance. Leonardo da Vinci is an icon for the synthesis of the multiple intelligences, scientific, emotional and artistic, identified by educator Howard

Gardiner in *Frames of mind*. Author Dan Brown's *The da Vinci code* is the best-selling book this century. As well as its adventure story, that book carries a powerful message on how we know what we think we know: how we construct our reality from the accounts we accept of the past.

There is plenty of evidence that synthesis is a successful tool for change. Individuals function most effectively when the parts of their life form a cohesive whole. Far-sighted managers strive to maximise their organisations' combined social, economic and environmental resources. Educators seek to inspire their students through cross-fertilisation between the arts and sciences. Sound public policy brings together health, education, community services and finance. Emergent fields such as computing design and transdisciplinary research are developing modes of synthesis that cross all knowledge boundaries.

Initiatives supporting collective thinking are well under way. They include Learning Circles, World Café, Laboratory for Social Intervention, Cognitive Justice Systems, Open Space Technology, National Coalition for Dialogue and Deliberation, Dynamic Facilitation, Deep Dialogue, Appreciative Inquiry, Critical Systems Thinking, Adaptive Management, Global Leadership Initiative, Governing the Commons, Pattern Languages and many more. These are not the products of impractical dreamers, but the output of some of the most influential creative and scientific minds. They are all readily accessible through the Internet search engines.

Whatever the area of application, and whatever the choice of contributors, the actual collective knowledge synthesis occurs at the time of making of a decision. In the field studies of successful sustainability initiatives that inform this book, lasting and constructive collective decisions were found to have a consistent set of characteristics. They were:

- future-directed: working towards a shared ideal, rather than tied to repairing past mistakes;
- collective: combining individual, community, specialist, organisational and holistic contributions on equal terms;
- open-ended: providing continual collective learning opportunities that involve all the contributors;
- a positive sum game: recognising that a collective decision adds value both to the individual contributors and to the whole enterprise; and
- rich with ambiguity and paradox: welcoming these essential elements of change-oriented decision-making that maximise potential.

Everyday experience confirms that effective decisions reconcile individual, community, specialist and organisational experiences. Check this out in your own life. In making important decisions, you find yourself drawing on the different capacities that contribute to your personal life, your community interactions, your area of special skill, your employing organisation, and to moments of creativity. In the end, any decision involves you reconciling them all, either knowingly or more often, just automatically. Try the exercise in box 1.2. When the contributions oppose one another, acute stress can be the result.

We are constantly amazed at the heights reached by the synthesis of these abilities in Leonardo da Vinci and his Renaissance peers. Yet our analytic tradition continues to mislead us into concentrating on their level of excellence in only one field at a time. Leonardo is considered a painter, Nicola Machiavelli a political scientist, Thomas More a philosopher. But the point is not that they reached such heights in each of these capabilities separately. It is that they brought all these capacities together in everything they did. It was the synergy between them that gave us the reflective notebooks of Leonardo, the integrated allegories of Michelangelo, the eloquent writing of Machiavelli and the utopia of More. During the Renaissance, individuals and their society developed this rich diversity as a cohesive whole.

Sadly, the social divisions of our time lead us to treat what could be elements of collective decision-making as if they are separate too. Not surprisingly, people then act separately, to the point of conflicts of interest. In Western society today, the two main contributors to any significant decision – specialist research and organisational agendas – vie to be the leaders in a hierarchy of knowledge. And individual, community and holistic contributions to knowledge are devalued. They may even be dismissed entirely. There is need for frameworks and tools that include all these ways of thinking in a creative synergy.

How shall we get there?

Leonardo's vision offers strategies able to generate collective thinking that will lead to collective action. The book takes a practical approach to collective decision-making, addressing synthesis from the starting points of an individual, a community, a profession, and an organisation in turn. Hopefully, it will demonstrate how the characteristics of a successful synthesis pose a powerful vehicle for positive change.

The chapter after this explores contributions to skills of synthesis since the Renaissance, and develops the idea of synoptic thinking. This leads on to a review of the ingredients of a constructive synthesis (chapter 3), and then to a suite of tools for arriving at synthesis-based decisions from different starting points (chapter 4).

The reader may open the book from their personal interest in integrating the complex issues of sustainability. It may be as someone responsible for teaching others to be collective thinkers. It may be in order to fulfill the role of expert synthesist in a profession or organisation. It may be as an agent of whole-of-community change, or as a holistic thinker. In each case, they will find in the chapters 5 to 9 their own point of entry to a collective synthesis. But only making use of all the synthesis tools described in chapters 5 to 9 will equip them fully as promoters of collective thinking. Chapter 10 proposes a collective decision-making system involving all the knowledge cultures on equal terms.

Figure 1.2 Leonardo da Vinci. *The Mona Lisa*. Paris, Louvre

© Photo SCALA

In examples to be found throughout the book, the practitioners of synthesis find that working collectively is demanding, but personally satisfying and mutually rewarding. But they also report feeling at a loss when they first decide to sponsor collective thinking. The accompanying changes include shifts in identity, career paths, and even notions of truth. There are, however, many successful ways to travel across the new landscape, which we explore in the chapters that follow. The reader can find an entry point by following the exercise in box 1.2.

The pathway is neither simple, nor easy. But there are precedents. The enigmatic smile of Mona Lisa lingers in our collective minds, across generations and across cultures. The unity of Leonardo's design gives the observer a deep understanding of the essence of his complex subject, in a way that resonates with all of us. At the same time, his experience in anatomical dissection and his observations of natural systems enable him to represent accurately the finest details of both person and landscape. Yet dissection was then punishable by death, and landscape had seldom been included in the picture at all.

Another collective thinker, Henry Nix, director of a think tank on issues linking environment and society, writes: "*We have oceans of data, rivers of information, and some pools of knowledge - but very few drops of wisdom.*" The journey we are about to take searches for some of those drops, through the fresh understanding that combining pools of knowledge can bring to the whole. With the emergence of chaos and complexity theories, there is a new spirit of reflection on the unifying power of whole systems, which we also explore. Leonardo proposed in his notebooks "*the part always has a tendency to unite with the whole, in order to escape its imperfection*". We have identified where we are now, lost in a fragmented world. As Leonardo predicts, we will also find a strong

Box 1.2

DECISION-MAKING: TESTING FOR PARTS AND WHOLES

As a test of the propositions in this chapter, identify a successful innovation in your own area of work or life, something that has made a difference. This could be a behaviour change, a technical improvement, or a management decision - for example, a decision to provide childcare in a workplace.

In the childcare example, it is likely that the change is attributed to feminist and community pressure. In practice, there will be social, financial and environmental costs and benefits for everyone concerned. Some change in the system will have brought about some new alignment between the decision-making groups. This change will have tipped the balance towards the actual decision.

A different individual may have come to hold the power, a new study may have identified greater output from workers when childcare is available. There may have been a shortage of workers for the organisation, or a shift in perception of childcare from a personal to a social benefit. Nevertheless, all of these ways of thinking will have contributed to the outcome. It is likely that only one group gets the credit. This will mean that the lesson that change needs all the decision-making groups together has not been learned for the future.

Consider the innovation you have chosen from your own experience. Find the personal, local community, expert, organisational and holistic influences that led to that decision. In almost every case, the full set of influences will have been involved, but not acknowledged, with the innovation being attributed to one sector, and one influence alone. In fact, the successful innovation will have been some form of reconciliation between all the decision-making groups.

current of thought going in the opposite direction, looking for the whole. In the next chapter we will identify other mentors who can lead the way.

2 Utopia revisited

"The diversity of our opinions does not arise from some being endowed with a greater share of reason than others, but solely from this, that we conduct our thoughts along different ways, and do not fix our attention on the same objects ... To be possessed of a vigorous mind is not enough; the prime requisite is rightly to apply it."

Descartes, R. 1637 *A discourse on method*

Figure 2.1
Leonardo da Vinci 1473 *Tuscan landscape* (his earliest known drawing). Florence, Uffizi.

Every era has its signature journey in response to the interaction between the times, the people and the place. Pathfinders of an era define their times and open the way for their society to reach its potential. Leonardo da Vinci's early drawings with their synthesis of scientific, social, artistic and personal observation, helped spark the inventiveness of the Renaissance. Then philosopher-mathematician Rene Descartes argued for using individual reason rather than artistic insight and religious instructions to interpret the way the world works. This produced the 400-year dominance of scientific analysis as the proper basis for thought. An ideal world was one that ran like clockwork. Albert Einstein introduced us to an understanding of the connectedness throughout the universe. He has helped turn our minds towards the interactive dynamic systems that go to make up a self-supporting whole.

Where are we now?

The swinging pendulum

By the end of the 20th century, people were seriously questioning the future direction of the technology-controlled world that had begun with the industrial revolution. There was a general climate of concern for a more united and sustainable world, a world that could deliver on the promises of democracy, social justice and a sustainable environment. However, those willing to act on the dire warnings of environmental and social collapse that threatened to derail this ideal were faced with the widespread disbelief of the Cassandra effect.

In ancient Greece, the prophet Cassandra had the gift of foretelling the future, but the curse that no one would believe her. Alan AtKisson, a Swedish-American turned advocate for global sustainability, states his case in *Believing Cassandra*. He beseeches us to accept the well-founded forecasts that our planet's oil and fresh water supplies have less than 30 years to run, and to act on that knowledge. He pleads for the figures to be believed so that we can ask ourselves afresh 'How shall we live?'. He notes that prediction is not destiny, and argues that we need to be fully informed about the direction the world is taking, in order to change it.

In *One world: the ethics of globalisation*, Yale University philosopher Peter Singer offers a measured assessment of the state of the world in 2002. He defines humanity's 'one world' as the combined resources of environment, economy, law and community. Singer concludes that industrialised countries' exploitation of the natural resources of the developing world has resulted in changes in the global climate. The physical and economic disruption reverberates around the world, returning in a feedback loop to affect industrialised countries. International regulatory systems such as the United Nations and the International Court lack the confidence of the world's nation states. The World Trade Organisation divides along the lines of resource-rich and resource-poor. Civil society and community struggle to find a place within these clashes of Titans.

Singer lays the combined scenario before us as a choice between dystopian thinking, that is, accepting this negative approach to the future, and utopian thinking, acting in hope of changing the world for the better. It could be a difficult choice unless we accept AtKisson's advice to take account of both. The synoptic response is to bring together all the information, both despairing and hopeful, as a baseline from which to take action towards that ideal.

You would not expect utopian thinking from the World Bank, the United Nations programs for the Environment and for Human Development, or the Organisation for Economic Cooperation and Development. These institutions are at the very core of the analytic and specialised mode of constructing knowledge. They are thus responsible for many of the compartmentalised decisions that have led to the divided world. Yet the advice they do give, to learn to adopt synthesis as the essential basis for working towards a sustainable environment, embodies the utopian ideals of AtKisson and Singer. This means turning towards utopia as an attainable goal, not an impractical dream.

American philosopher Thomas Berry, in *The great work*, calls this time in history a moment of grace. He compares the present challenge with other major breakthroughs of human invention such as controlling fire, using the wheel and harnessing fossil energy. In our time, the combination of the information revolution and unregulated globalised flows of resources are changing the world. Taking advantage of the opportunities that we have to change it in a positive direction requires a vision of where we could be going, rather than being driven by the dread of what we need to escape. Berry suggests that a sustainable humane future is still possible, if we learn to harness the synergy of bringing together all we know.

Fortunately, if we follow the AtKisson-Singer-Berry agenda we do not have to start from scratch. The powerful tools of the scientific era are still at our disposal if we can focus on how to redeploy them towards a shared goal. There is already a language of synthesis (box 2.1) and, as we shall find, a practice of synthesis, however much these may have been masked by the ruling preference for the linear logic of analysis. It was the tradition of linear reasoning that constructed the story that human understanding of the universe advanced in regular steps from the naïve naturalism of the Druids to the breaking of the atom. This idea of steady intellectual development could not be more mistaken.

In his 2001 book, *Truth: a history and a guide for the perplexed*, Oxford historian of ideas Felipe Fernandez-Armesto documents the changing styles of collective wisdom that have come and gone over the ages. How we know what we know has changed significantly in each historic era. An era may break new ground and yet have the advances disappear a generation later. In other cases, a new way of thinking may survive centuries of suppression. The

Box 2.1
THE LANGUAGE OF SYNTHESIS

Collaboration:	individuals work as a team for a common cause
Collective:	individuals agree to work together to common rules
Cooperation:	individuals or groups assist one another to a common goal
Coordination:	linking the working parts of a human or biophysical system
Hierarchy:	a graded set of unequal parts
Holarchy:	a connected system of equivalent wholes
Holism:	seeking the core or essence of a system or an idea (alternatively applied to understanding the whole of the whole)
Holocentric:	direction towards the core or essence of a system or idea
Integration:	forming a unified whole from individual parts
Synergy:	interactions between parts that enhances both parts and whole
Synoptic:	insight generated from different perspectives on the one theme
Systemic:	insight generated by noting links between the parts of a whole
Synthesis:	a fresh whole formed from interacting parts

Roman conquest of the Greek civilisation and yet the strong continuing influence of Greek philosophers on Western thought is but one example.

The American journalist-philosopher, Jeremy Campbell, in *The liar's tale: a history of falsehood*, also 2001, traces the way that even the type of lie we tell reveals the way we think about the truth. Mediaeval surgeons reported the missing rib in males that their belief system assured them had been taken from Adam to form Eve. It seems that lies, like truth, are factors of time, place and culture. Writers from Jurgen Habermas to Michel Foucault confirm that our own time is one in which we have begun to reflect on ideas about truth and knowledge, and reconsider our ways of interpreting reality.

American historian of ideas Stephen Toulmin finds that the construction of knowledge has regularly swung between priorities for analysis and synthesis (figure 2.2). In the Western cultural tradition, the swing of the pendulum towards synthesis in thought and action includes the peaks of early Greek civilisation of around 500BC, the Renaissance around the 1500s, and the Information Era of a globalised world during the early 2000s. These times have alternated with eras of religious dominance, and more recently with the dominance of the search for objectivity. The swings are by no means uniform, and smaller swings happen along the way. Toulmin concludes that analysis and synthesis remain essential components of all ways of thinking, but the emphasis changes according to the demands of the time.

In times when analysis dominates, Toulmin suggests the questions being asked become more and more specialised and fragmented. They eventually become distanced from the larger, more urgent questions of the time. At this stage, the priority moves to addressing the socially pressing questions rather than the discipline-oriented ones. Understanding the essence of an issue becomes more important than the finer details of

Figure 2.2 Swings between analysis and synthesis (after Toulmin 1977)

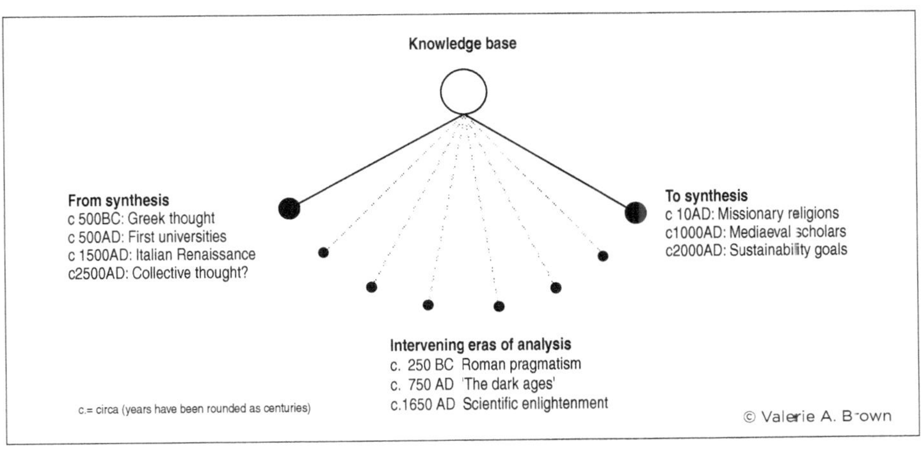

what is already known. Singer's overview of the state of the world, above, is an example of a synthesis offering a deeper understanding than the detailed economic and environmental reviews. Eventually the knowledge gained about the whole suggests new directions. Following those directions generates new detailed questions, and so the swing of the social pendulum returns to a preference for analysis.

Not only philosophers but creative writers and scientists have also confirmed Toulmin's conclusions on the 20th century as the turning point towards synthetic thinking. In the mid-20th century, the Nobel Prize for Literature had been awarded to a novel that explored ideas about the collective construction of knowledge. Hermann Hesse's tale, *The glass bead game*, is the story of a game played by a group of the world's most super-intelligent men. They have committed their lives to bringing together all that has ever been known in the world, in order for them to make wise decisions towards a more perfect world.

One player, Joseph Knecht the Magister Ludi (master player), ultimately completes the game. He finally knows how everything is connected to everything else. He has successfully discovered *"the same eternal idea that was the basis of every movement of the mind towards the ideal goal of every Platonic Academy, of any congregation of a spiritual elite, every attempt to approximate the exact and the free sciences, every attempt to reconcile science and art, and the art of science and religion"*. In the end, however, he can put none of it into practice because there is no language or shared experience which would allow him to share his discovery with others.

The glass bead game is still being played worldwide. There are at least ten variants on the WorldWide Web, each with its committed band of followers. The work on pattern languages which forms the basis for chapter 9 on holistic thinking can be thought of as a variant on the glass bead game.

In science, the physicist Fritjof Capra published *The turning point: science, society, and the rising culture* in 1983. The book evaluates the state of the physical and social sciences, confirming Toulmin's prediction that the specialisations had become so fragmented as to generate a turning point towards synthesis. Capra has followed this diagnosis by a further book on the elements of a synthesis between the social and physical worlds *The web of life*. His key elements for a lasting synthesis are differentiation of the parts, self-organisation of the whole and transparent communication about the system. These are central themes throughout this book.

A signpost for guiding social change is frequently attributed to Albert Einstein: *"It is not possible to find the solution to a complex problem from within the system of thought that created it"*. H. J. Rittel and his colleagues in the seventies were saying much the same thing when they coined the phrase ' a wicked problem' for problems that lie outside the capacity of the current society to resolve. The limitations of our present system of divided thought and practice fit the definition of wicked problems. The solution is to follow those who have tried to understand the whole of a system, and have found the courage to think outside it.

Where do we want to go?

Making the jump to collective thinking

Many logical reasons are put forward to stop us meeting the challenge of synthesis as a primary way of thinking. One is the structure of our large brain, with its twin lobes housing the separate skills of analysis and synthesis. Another is that the division of knowledge into specialisations has led to the triumphs of the past technological era, so we disturb that at our peril. Yet another is the belief that if we change the ways in which our divided forms of knowledge have been constructed in order to reunite them, the changes will alter the social system itself.

Each of those arguments has a measure of truth, and yet for each there is a positive response. For the first, the human brain's unique capacity lies in the ability to hone and integrate the separate skills of analysis and synthesis. Second, our technological inventions have altered our planet so comprehensively that the very success of the emphasis on analysis has led to the need to apply integrating tools to the outcome. Third, linking community, organisational and professional thinking and practice will indeed change the social structures currently built around their divisions, and that would be no bad thing.

An example of the ways in which the frictions between the personal, community, specialised, organisational and holistic aspects of our everyday lives evolve into a self-reinforcing system is in box 2.2. It is difficult, if not impossible, for change to occur unless these multiple dimensions are recognised and brought together in some way.

As in box 2.2, the strength of the self-reinforcing social system means that changing one element requires transformation of the whole system. In Europe, attempts to increase daylight working hours in an effort to lower energy consumption from lighting and heating in order to reduce global warming, met with little or no success. On the other hand, great enthusiasm greeted a later French decision to reduce working hours in order to increase family and community time together. Although it led to higher work output and social morale, the move unraveled four years later because of larger economic and international organisational pressures.

When the decision-making groups accept that they have different rules for constructing reality, the whole game changes. The physicist Niels Bohr gave us *complementarity*, the idea that two different interpretations of the world can be true at the same time. Light is made up of both waves and particles, depending on the rules applied by the observer. From Bohr's pupil, the German physicist Werner Heisenberg, we learnt the inevitability

BOX 2.2

TESTING FOR SYNTHESIS: THE WESTERN WORKING DAY

Our biological clock is permanently tuned to each calender day, beginning as the sun rises and ending as it sets. Our individual body rhythms have adapted that clock to the way in which we live our lives in a given society. Our community's customs override our biological and individual rhythms, setting the times when we are all expected to work and rest. Communities around the world set their clocks with reference to global time, checked by a nuclear device in Paris. Technology, therefore, allows us to time our day as part of a global population, across continents, working independently of dawn or dusk, or the seasons.

Our organisations and governments instruct us on the pace at which we should work. They tell us when to work, how long, with whom, and for whom. There are legal restrictions on the timing and conditions of paid work, but little formal consideration of the time required to meet personal, neighbourly and social obligations.

The timetable we keep is kept in place through a synthesis between our biological messages, social customs, civil rights, market forces and the local climate. Think of your own working day and its personal, community and professional demands. What would you need to change to bring it closer to your own preferred day?

What would be the effect on your relationships with your family and your community, and your organisation, if you decided to change your allocation of your personal time?

of the *uncertainty principle*. Any observations of social or physical reality will be influenced by the means of observation, and the position of the observer.

It follows that we can never know anything for certain or forever. There is only a particular human perspective at a certain time. A recent advance in our understanding of the physical world comes from Ilya Prigogine the chemist and Benoit Mandelbrot the mathematician. From very different experiments, namely heating liquids and mathematical modeling, each demonstrated the internally imposed order that is present even in the most complex chaotic system. The smallest change in any system can trigger its self-organising properties. Hence Edward Lorenz's famous question, based on studies of how air moves around in the atmosphere: *"Can a butterfly wing flapping in Brazil create a tornado in Texas?"*. The answer is yes, and the so-called butterfly effect has been labeled deterministic chaos.

An idealised plan for a collective communal life was Thomas More's 16th century Utopia, a detailed design for constructing a perfect city. It hardly deserves its reputation for being impractical. He writes of Utopia under the headings *Of Their Towns, Of Their Magistrates, Of Their Trades, Manner of Life, Of Their Traffic, Of the Travelling of the Utopians, Of Their Slaves, Of Their Marriages, Of Their Military Discipline* and *Of Their Religions*. He also notes that the biggest issue is to avoid becoming bogged down in the problems but to rise above them:

"There were nations, towns, and cities, that had not only mutual commerce among themselves, and with their neighbours, but traded both by sea and land, to very remote countries. We asked him many questions concerning all these things, to which he answered very willingly; only we made no inquiries after monsters, than which nothing is more common; for everywhere one may hear of ravenous dogs and wolves, and cruel man-eaters; but it is not so easy to find States that are well and wisely governed."

Thomas More, 1530, *Utopia. Book 1.* p1.

How can we get there?

Creating a collage, not a jigsaw puzzle

A collective approach falls within the meaning of Descartes' advice at the head of this chapter. We have already noted that his proposal that the most important thing for humans to do is to rightly apply their minds applies equally to analysis and synthesis. Yet for some centuries it has been assumed to refer to analysis. Figure 2.3 offers a simple test case of the wide range of possibilities in determining wholes. Anyone looking at the figure in reductionist mode is able to find the parts, the 26 squares, without much trouble. But you might like to consider how many wholes – that is, collection of parts – you can find in the figure before you read the next paragraph.

Figure 2.3 Testing the relationships between parts and wholes*

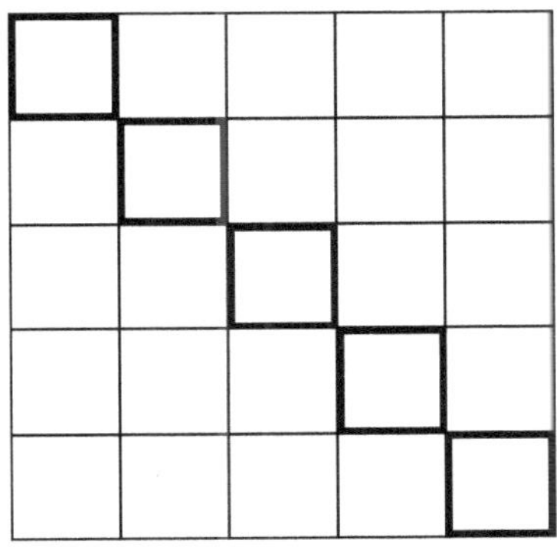

* How many wholes does this figure contain?

© Valerie A. Brown

When we change the purpose of the exercise from how many parts to how many wholes you can find, there are well over 500: 480 sets of pairs, 20 sets of triads, eight sets of four, seven diagonals and two concentric sets. Each set of wholes conforms to a rule that sets it apart as a whole. The resulting pattern may represent a frame (if you fill in the outside squares), a correlation (the main diagonals), a proportion such as 25% (fill in one-fifth of the squares) or a nested set (each ring of squares moving from the outside in). Each pattern is a whole in itself and part of the larger whole at the same time. The rules for identifying wholes call for us to celebrate complexity and difference, not attempt to eliminate them.

Words for the fresh insight generated through bringing together sets of wholes include 'synergistic' and 'synoptic'. *Synergy*, as we have discussed, is generated in cases where the parts of a whole interact with one another in a way that makes them more effective. *Synoptic* describes the case where separate wholes contribute as separate units to the creation of further wholes, as in our simple grid in figure 2.3.

Examples of synoptic sets are listed in box 2.3. The most familiar application of synoptic thinking is the daily television weather chart. Forecasting the weather, long a pet subject for ridicule, has become more reliable and more useful since being presented in the form of a synoptic chart, pulling together changes in temperature and barometric pressure, humidity, past weather patterns, local topography and global weather cycles.

Synoptic treatments of their material have allowed many writers to generate unexpected insights on their core themes. One of the most influential has been the synoptic gospels, parallel accounts of the life of Christ by three of his followers. A synoptic suite of novels on the moment of partition of Pakistan from India in 1947, Peter Scott's *Raj Quartet*, retells the experience of an ageing missionary, a young Hindu activist, a respected Moslem lawyer and a British army officer. More recently, the *Matrix* films vividly document parallel realities. In each example, the separate accounts of the whole differ from each perspective, but each remains valid in its own right, while contributing to a greater understanding of the whole.

Synoptic interpretations have even been creeping into our science knowledge base, although not without a struggle. One member of the Nobel prize-winning team who discovered the double helix shape of genetic DNA was the American James Watson. In his autobiography *The double helix* he attributed his discovery to a successful leap of the imagination based on parallel contributions from an innovative biochemist, a painstaking radiographer, and an experienced immunologist. He considered himself to be a scientific entrepreneur. Scientific reviewers reacted with shock to Watson's account of the synoptic nature of the discovery. Other scientists preferred the more traditional image of scientific discovery, of collaborating scientists slowly accumulating carefully verified facts.

The same sense of shock greeted a spectacular piece of synoptic thinking by James Lovelock, a British industrial chemist, when he came to the conclusion that the planet Earth acts as a single living organism. He linked the planet's life-supporting oxygen, water and carbon dioxide cycles to the small localised chemical reactions taking place in oceans and soils. He was able to demonstrate that the stability of the planet is maintained through the synergy of this vast network of feedback interactions. Lovelock was convinced by a writer to name the planet's system Gaia, after the Greek goddess of the earth. This title proved his professional undoing. It led to his scientific colleagues initially rejecting his work as a metaphysical, not scientific, explanation. In his 2000 autobiography *Homage to Gaia* he describes the anger that he dared to link two aspects of reality usually kept separate – the world of ideas and the biophysical world. He also points out that two decades later, the same scientists were funded to expand our understanding of Gaia.

Leonardo da Vinci, another practitioner of synthesis, united parallel strands of knowledge whenever he produced a painting, a machine or a sculpture. His dissections of the human body informed his work on his human figures, at a time when dissecting could be punished by death. His accurate and aesthetically satisfying drawings of mechanical devices and natural objects drew the eye to a new understanding of physics. Leonardo explicitly acknowledged the different worlds of mathematics, physics, aesthetics, personal experience and a community's common sense. He writes of deliberately thinking synoptically, in order to produce the whole: "*Although the painter's eye sees but the surface of things, it must in rendering the surface discern and interpret the organic structure that lies beneath. The representation of the human figure entails a knowledge of its anatomy and its proportions. Landscapes entail a study of the morphology of plants, of the formation of the earth, of the movements of water and wind*".

After considering the advice of the synthesists quoted above, and the examples in box 2.3, it becomes clear that synthesis involves entering new territory. The options for breaking new ground can be compared with producing an original work of art, completing a jigsaw puzzle, or creating a collage. In an original work of art, new knowledge emerges from within a creator's head and hands. In completing a jigsaw puzzle, there is an existing picture to work towards, and the pieces either fit or don't fit a pre-existing pattern. Creating a collage is both creative and synoptic, in that the separate contributions remain distinct, while transforming the way both creators and observers understand the whole.

Compare the Picasso collage of three musicians in figure 2.4 with a straight portrait of the three musicians, and then with a jigsaw of the collage. Unlike the other two, the collage requires reflection, imagination and commitment to an idea, on behalf of both artist and viewer. It uses ambiguity and uncertainty to lead the observer to create

Box 2.3
GOSPELS TO GOOGLE:
EXAMPLES OF SYNOPTIC PRACTICE

Synoptic syntheses in which the parts are independent of one another, but each contributes to the understanding and strength of the whole:

- The gospels of the St James Bible
- The weather report
- Democratic voting system
- A judge's summing up
- A multicultural society
- Google search engine
- A Shakespearian play

a new whole, one not realised in quite that way before.

Figure 2.4 Pablo Picasso *Three musicians* (collage) New York, Museum of Modern Art.

The concern to make better use of the human capacity for synthesis, both within each individual mind, and between diverse minds, comes from many directions. Collective thinking can take many forms, some that we wish to harness and some we learn to avoid.

Carl Jung's idea that all societies are held together by a shared cultural unconscious, a biological given, forms the basis for a major school of psychiatry. Teilhard de Chardin in his *The phenomenon of man* in 1955 proposed an evolutionary trend to a Noosphere, in which all minds would meet and contribute to each other's knowledge. Tools for this mutual contribution have emerged in a cyberspace he could not have even imagined. Outpourings of collective sympathy have responded to victims of famine and natural disasters in distant parts of the world. A popular song sweeps across whole populations. Collective thinking already exists.

Unfortunately, collective thinking is sometimes mistaken for the mindless conformity of George Orwell's 1984 satire *Animal farm* and the inhuman world of obedience to an authoritarian regime, as in Arthur Koestler's description of the Stalinist regime *Darkness at noon*. 'Group think' and cases of mob hysteria could also be ascribed to the Noosphere. This would be seriously misleading. Collective thinking rests on the freedom of individuals to think for themselves, not the power of a leader or a manifesto. The success of a synthesis arises from the quality of the individual contributions, whether it is the chemical reactions of Lovelock's Gaia or the painstaking research undertaken by Watson's colleagues on DNA.

The management world believes collective thinking can be taught. In a 2004 study by Andrews and Tyson, the world's business leaders considered that the most important goal for management courses was for future business leaders to learn how to synthesise diverse knowledge domains while accepting conditions of ambiguity and uncertainty. The American Jesuit philosopher Thomas Berry sets this capacity as a condition of human survival. He writes that the 'great work' of humanity is to work collectively to ensure a secure future for all living things. The economist Marilyn Waring proposes a collaborative, constructive economy in the place of the competitive marketplace.

There is no lack of leadership, ideas and strategies that can be explored in establishing a capacity for future-oriented collective thinking. The synoptic approach described above provides the initial step. Chapter 4 goes further, to construct a collective tool kit for arriving at constructive syntheses. But before we can make use of these tools we need to better understand the relationships between the parts and the whole in the all-pervasive Western culture. Chapter 3 explores the current divisions of knowledge in Western thinking. It examines what it would take to link key decision-makers in constructive collaboration, while respecting their existing knowledge bases. This brings into focus the parallel versions of reality we have been accepting as knowledge, and some of the new ideas we may have been dismissing as ignorance.

3 Parallel worlds

"There where it is we do not need the wall.
He is all pine and I am all apple-orchard.
My apple-trees will never across
And eat the cones under his pines, I tell him.

He only says 'Good fences make good neighbours'.
Why do they make good neighbours? Isn't it
Where there are cows? But here there are no cows.
Before I built a wall I'd ask to know
What I was walling in or walling out,
And to whom I was like to give offence."

Robert Frost, 1972, from *Mending Wall*

A joke about the European Union runs thus: in a United Europe, the cooks should be French, the police British, the financiers Swiss, the administrators German and the lovers Italian. This is only half a joke. Of course, it is not possible to pick out the eyes out of a culture. Within the boundaries to those countries the valued expertise has taken centuries to build. On the other hand, we can take seriously the idea that if each culture could contribute its strength to the whole while still maintaining their own distinctiveness, the whole of the new Europe would benefit. In a collective knowledge culture, can we therefore expect individuals to contribute their commitment, communities their visions, specialists their accuracy, organisations their political skills and holists their creativity?

BOX 3.1
COLLECTIVE KNOWLEDGE ... ITS FORMS AND PROPONENTS

• The nature of order	Alexander 2005
• Modes of reason	Nanno 2003
• Western knowledge cultures	Brown 2002
• Complex Adaptive Systems	Costanza 2002
• Sustainability science	US Research Institutes 1999
• Communities of practice	Wenger 1998
• Ways of knowing	Belenky et al 1997
• Triple Bottom Line (TBL)	Ecological economics c.1995
• Sustainable Cities	European Union 1996
• Healthy cities	World Health Organisation 1986
• Games of truth	Michel Foucault 1975
• Situated logics	Toulmin 1958

Sources: see bibliography

Where are we now?

Multiple realities

Many writers have attempted to reconnect the modes of knowledge that in Western culture have been long divided. Box 3.1 contains a list of such syntheses in a range of settings. They include synthesis of abilities within each individual:- ways of knowing, forms of learning:- in whole-of community settings:- cities, environments; among the professions:- transdisciplinarity, communities of practice; management practices:- the triple bottom line linking social, economic and environmental resources, complex adaptive systems; social roles:- games of truth, situated logics; lines of argument between different interests:- modes of reason; and holistic urban design:- pattern languages.

The task in this chapter is not to argue for or against any one of these frameworks. It is to consider how they can usefully be interconnected within a common system. Collective decision-making – in a synoptic approach – takes account of each of the contributing sectors, rather than attempting to choose between them. This is not as difficult as it may first appear. The solution lies in ceasing to concentrate on one abstracted aspect of reality at a time, as in literature, mathematics or biology. This is what our traditional education system has trained us to do. The basis for collective decision-making is achieved through looking out the window and observing events in the real world in real time. All significant events involve interactions among individuals, communities, specialists, organisations and holistic thinkers. Cooperation in an appropriate collective response depends on mutual understanding among these players.

As an antidote to fragmented and divisive constructions of reality, this solution may seem simplistic. Yet it reaches to the heart of the system that is currently supported by fragmented decision-making. In practice, the fragmentation meets H.J Rittell's 1960s definition of a wicked problem, one that lies outside the capacity of the current society to resolve. Within Western societies, a hierarchy of knowledge privileges academic disciplines and dominates any attempt at a synoptic approach. Collective decisions therefore require a serious reorientation of our present construction of knowledge. Before that can be done, it is necessary to examine how we arrive at that knowledge, and how it is currently divided.

Sociologists Peter Berger and Thomas Luckmann in *The Social Construction of Reality* document how all knowledge originates in a human head, where it is shaped by the cultural and physical context. Schools, professions and the media do not transmit knowledge, but information. The teaching/learning strategies they use lead to it being absorbed (or not) into each individual's own construction of reality. Contrary to popular usage, knowledge is not the inert information stored in libraries and textbooks, but a dynamic interaction between people, time and place.

The Oxford *Dictionary of Philosophy* defines 'knowledge' as justified true belief. Our knowledge is founded on a combination of a respected authority and our individual experience. In the Western tradition, the source of that authority has changed markedly over the years. Fernandez-Armesto traces its path in *Truth: a history and a guide for the perplexed.* Broadly speaking, the idea of truth in Western culture goes back to the humanist tradition of the ancient Greeks, which assumed that each individual could achieve wisdom (although this did not include the slaves). This was followed by respect for practical knowledge by the Romans, and then by the reliance on religious texts of the Middle Ages. This moved on to faith in the creative human spirit during the 15th century Renaissance. Since the seventeenth century, the accepted source of valid knowledge has been the analytic, objective observations that the world hailed as the scientific Enlightenment.

There are a variety of initiatives that suggest a move to a construction of reality based on synthesis. We have mentioned Lovelock's Gaia, and Berry's *The great work*. Various other candidates include the Information Highway, Globalisation, and the New Renaissance. *Panarchy*, with a subtitle *understanding systems of transformation in people and nature*, is the term environmental scientists Gunderson and Hollings propose for working for sustainable development. Sociologists have attached the label 'post-post-modern' to the ideas of writers such as Michel Foucault, the French structuralist thinker. Foucault in *The order of things* conceives knowledge as the source of power and control, within a system in which all social and physical events are interconnected, influencing each other.

A timeless metaphor for the way in which humans construct reality is found in Plato's parable of the cave. Plato suggested that we see physical events as if we are slaves chained in a dark cave, with our faces to the walls. All we can make out of the world is from the shadows cast on the wall by the torches held by people behind us. Nevertheless, we strive to make the best interpretation of reality we can out of what our senses and our tools for thought can tell us. This will, however, never actually match the external reality.

The parable continues with the prisoners emerging from the cave to the outside world. They cling to the world as constructed in the cave, in the face of the overwhelming evidence of a different reality. Examples of such denial of a new experience are well established. The psychologist Milgram found that, faced with new evidence, one-third of his subjects held fast to their previous belief, one-third were prepared to change if authority convinced them, and one-third accepted the new reality from their own review of the evidence. A 21st century example has been the slow acceptance, first by the public and then by governments, of the evidence of global warming.

In 1970, American sociologist Thomas Kuhn in *The structure of scientific revolutions* established that, in spite of the rhetoric of the Enlightenment, knowledge breakthroughs do not arise in orderly sequence from careful, detailed analysis of existing evidence. From a review of research practice he demonstrated that new ideas arise from 'revolutionaries' working outside normal practice. He also found that, initially, 'normal' practitioners reject the new ideas. Eventually enough light is cast on the existing frameworks from the new forces for normal practice to be re-directed to accept the new framework. The question arises: is collective knowledge normal or revolutionary, in Kuhnian terms?

Figure 3.1 contains an overview of the knowledge platforms of the different decision-making sectors that need to be brought together in order to solve wicked problems. Following the work of Kuhn and his successors such as Imre Lakatos and Bruno Latour, each knowledge platform can be shown to have all the dimensions of a full construc-

Figure 3.1 The knowledge cultures of Western decision-making (Brown 2001)

KNOWLEDGE CULTURES AND CONTENT	STRUCTURE	SOURCES OF EVIDENCE	ASSIGNMENT OF IGNORANCE
INDIVIDUAL KNOWLEDGE Lived experience, identity		Memory Learning style Five senses	Subjectivity Narrowness Vagueness
LOCAL KNOWLEDGE Shared lived experience of people and place		Stories Events Histories	Gossip Anecdote Inaccurate
SPECIALISED KNOWLEDGE Mono, multi and trans-disciplinarity, the professions		Case studies Measurements Observations	Jargon False centainty Narrowness
STRATEGIC KNOWLEDGE Administration, government, organisations, business	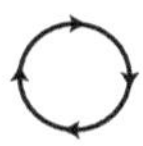	Agendas Alliances Networks	Deals Mates Corruption
HOLISTIC KNOWLEDGE Essence, core of a system		Synthesis Focus	Airy-fairy Impossible Ambiguous

tion of knowledge. Each has its own knowledge content, mode of collecting evidence, tests for truth, language, and way of rejecting the others. Left to themselves, each of these knowledge platforms has become sufficiently self-contained to be described as a separate knowledge culture, the term which will be used from now on. Each knowledge culture strongly defends its own version of 'truth' so that to put them together in synoptic fashion is to transform present practice.

Thinking individually

Every individual's knowledge is unique. Individuals build knowledge out of their own lived experience, shaped by their social and physical setting. We all tend to reject material that does not match our own version of reality, as a way of maintaining our own identity. The creation of each person's identity is a synthesis of the information on the world sent to him or her from their intellect, their physical experience of the world, and their ruling inspiration. United States science administrator Jack Nonna calls this last dimension the 'over-riding instructions' that determine the reaction to the other two. Mediaeval philosopher St Boniface refers to a full understanding as requiring the eye of the mind, the eye of the body and the eye of the spirit. The creative leap of imagination that brings all this together is perhaps the most uniquely human of all human capacities.

Each individual's identity is the outcome of their internal 'eyes' which interpret the influences of community values, authoritative evidence and organisational rules. Michael Polanyi, in his seminal *Personal knowledge*, distinguishes between explicit and tacit knowledge. Tacit knowledge is the knowledge we all draw on all the time, without being directly aware that we hold it. Individual knowledge is represented in figure 3.1 by the cloud of small dots. Strategies for an individual to draw on all their abilities, tacit and explicit, are discussed in chapter 5.

Our individual knowledge synthesis is expressed tacitly in many ways, all of the time. As an example, consider the clothes you are wearing. Our clothes reveal the style of our negotiation between our identity and our environment. Do you follow fashion? Or lead it? Imagine you are wearing the gear of the other gender, or the dress of another culture. Or less clothing than you usually wear. How would people respond? Your workmates? Your organisation? Your family? In extreme cases, even the local police? How would you respond to each of them? Would you feel confident in being yourself in any dress? A cultural historian, a biochemist, a sociologist, a psychologist or a dress designer would give different descriptions of your clothes, even if they had all made rigorously objective observations. But your choice of clothing is uniquely you, based on your own individual interpretation of reality. The synthesis is your own.

Thinking like a 'local'

Residents of towns or cities often claim that newcomers can never be a true 'local', even after they have spent many years in the area. How do communities become such exclusive clubs? Local communities are cemented together by a diverse body of local knowledge that is legitimised through shared experiences, memories and symbols of places and events. Each community member develops a sense of belonging to a people and a place, and holds a body of knowledge accessible only to other community members.

The body of local knowledge is largely inaccessible to outsiders. Observers who develop mutual trust with a community may gain access to sensitive and confidential local information, but such knowledge must always be second-hand. This applies to all those outside the community network, even when they are highly-skilled social researchers or experienced service agencies. Scientist Brian Wynne has documented particular cases where the local knowledge of events was more accurate than that of expert scientists. In his classic study of the radioactive fallout around the British Windscale nuclear plant, *Where the sheep may safely graze*, the farmers' commonsense observations proved more reliable than those of research scientists.

Local community knowledge is built up by people acting as residents and citizens, rather than their other roles as individuals, experts or employees of organisations. Each community is distinctive, but linked to others in a network in the local region, and across the nation. Hence local knowledge is represented as a continuous but changing wavy line in figure 3.1. We can talk of the global community of humankind, a community of practice among people who share a set of skills, or our own communities of friends and neighbours. Clifford Geertz, cultural anthropologist, writes of communities as cultures of commonsense: "*Not just using their eyes and ears, but using them collectively, judiciously and reflectively to understand their own locality*". Chapter 6 contains an account of the way symbols are used to construct this mutual understanding within every type of community.

Becoming a specialist

The strength of specialised knowledge lies in its firm focus on a given topic, within set rules of inquiry and established explanatory frameworks. Each framework, be it in law, literature, history, philosophy, geography or physics, services a distinct sub-set of practitioners and practices. Ever since the Enlightenment, the subsets have been proliferating. At Oxford University in the nineteenth century, philosophy included physics. Today we have analytic, environmental and moral philosophy, with a separate discipline of physics, itself divided into astrophysics, geophysics, and more.

Since the time of the scientific Enlightenment, each new discipline has tended to adopt the methods of empirical science. This includes the social sciences, and even the humanities. The scientific mind-set fosters a search for objectivity, simplicity and certainty in the phenomena under study. Critiques of the limitations of scientific method for addressing complex problems such as sustainable development have led to a broader perspective within the scientific community. Methods for managing complex science-related issues include the idea of *post-normal science* developed by Ravetz. This includes recognition of uncertainty, complexity and the role of values in the interpretation of reality. However, most of the nitiatives in science still retain their commitment to a monopoly of objective and reductionist inquiry, confirming Kuhn's proposition of the difficulty of revolutionary change.

Even after a generation of teaching the sociology of science and acknowledging the inescapable subjectivity in the construction of knowledge, there are still strong claims from universities that all reliable knowledge resides within the academic disciplines. It is not unusual to find requests for integration of evidence and for collaborative decision-making met solely from within the specialised disciplines and professions. The tests for truth in the personal, community and organisational knowledge cultures listed in figure 3.1 are dismissed as being subjective (which they are) and unreliable (which they are not).

Since there is no inherent connection between disciplinary frameworks, integrative frameworks on issues such as sustainable development are imposed from outside, as in the notions of *Gaia* and *Panarchy*. Specialised knowledge is represented as a disconnected ring of boxes in figure 3.1. Box 3.2 offers a case study of a synoptic process which successfully brought specialised disciplines, individual reflections, community engagement, and organisational power together with the capacity to act. In chapter 7, it is suggested that a set of shared commonsense questions can bridge the distance between the disciplines, and community, expert and organisational knowledge.

Organisational realities and strategic thinking

Effective responses to complex issues which challenge society call for an understanding of the political and administrative realities. Even the most accurate and reliable of scientific evidence is not enough by itself to move a community to take action. There also needs to be a sense of direction and the power to act. A sense of direction comes from individual and community knowledge cultures. The capacity to act comes from the organisational structures that go to make up the society, the hospitals, banks, industries, services and government agencies. In any type of organisation, knowledge is strategic, directed towards achieving organisational goals. Validation of the soundness of the knowledge is tested by progress towards the desired outcome. This is represented in figure 3.1 as a closed circle with directional arrows.

Organisations, including industry, govern-ment and non-government agencies, usually give priority to pursuing their own internal goals. Collective thinking and action on an issue such as sustainable development requires that the executives of influential organisations consider other players, including its own individual members and the surrounding community. One has only to consider the initial responses to the suggestion of reducing the use of fossil fuels to check climate change, to realise that collective thinking requires a marked reorganisation of current practice. Coal and oil companies and Western governments ridiculed the suggestion. Renewable energy interests, environmental scientists and conservationists welcomed it.

Organisational or strategic knowledge is discontinuous across geographic scales. The management of a small locality differs in form and function, not only in size, from that of a wider region, a nation, or the planet. Prime examples of how strategic knowledge has informed decision-making throughout the centuries are the Chinese *Art of strategy* by Sun-zhu, the *Analects of Confucius*, Machiavelli's *The prince* and the satirical television series *Yes, Minister*.

As for each of the other knowledge cultures, there is a 'normal' and a 'revolutionary' form of organisational knowledge. Examination of private and public organisations confirms that the 'normal' form for both industry and government is increasingly that of the private for-profit organisation. All types of organisation are constructed around a social purpose, whether that be competition or collaboration, self-serving or outward looking. The consequences of a 'revolutionary' change in organisational decision-making are explored in a holocentric framework in chapter 8.

Holism's creative leap

Holistic knowledge can be classified in

two contrasting ways: as the essence of the whole, or the whole of the whole. With the latter, holistic thinking is sometimes dismissed as a presumptuous attempt to know all about everything. Or it is applied to thinking systemically, to representing reality as a set of interconnected parts. The whole might be an administrative chart, a food chain, or a social system.

We use the word 'holistic' here in its original sense, as the rationale, purpose or essence of a dynamic whole. The word was coined as recently as 1931, when the South African political leader, Field Marshall Johan Smuts, described holism as 'the tendency, as within nature, to form dynamic self-defining systems of wholes'. From the Oxford English Dictionary: *"Holism: a term coined by General J.C. Smuts to designate the tendency in nature to produce wholes (i.e. bodies or organisms) from the ordered groupings of unit structures."* (Oxford English Dictionary 1971: p399)

Smuts himself was more imaginative than the dictionary implies. As its inventor, he found the need to coin the word to describe the strength of his commitment to the development of his drastically changing country. He used it to bring attention to aspects of existence that were being bypassed by the so-called reductionist specialist tradition, which took account of physical observations only. His ideas and ideals were based on identifying the connected, synergistic patterns that can be found throughout all natural systems.

Holistic knowledge is constructed through the creative leap - the Aha! that gives access to fresh understandings. Biodiversity is a holistic idea, entropy another, coined to identify the essence of complex systems that at the time had no existence in the language. The creative leap is often regarded as the task of the arts alone. Yet it is an essential part of all forms of knowledge construction, the step in thinking that brings all the others together. As synonyms for holism, Stephen Toulmin uses 'synthesis' in *Of Human Understanding*, and Isaiah Berlin uses 'cohesiveness' in *The Crooked Timber of Humanity*. Other interpretations of holism include harmony, the pattern that connects, and life itself. Each of these uses is discussed further in chapter 9. The structure of holistic knowledge represented in figure 3.1 is in the form of a star, or the focus of a collective knowledge system.

Knowledge and ignorance

It is not only the construction of knowledge that is important in arriving at collective decisions. The construction of ignorance is equally significant. Assigning ignorance is the method by which we define what is proper and improper to know and do. The Buddhist mantra 'There can be no theft without property; and no ignorance without knowledge' applies here.

An Australian psychologist, Michael Smithson, has drawn up a schema on the role of ignorance in the construction of knowledge. He distinguishes between labeling ignorance as error (you are mistaken), as irrelevance (you are talking nonsense), or as taboo (you are not allowed to even think about that). The first label, of being mistaken, at least admits the mistake is a variation, even if false, of familiar knowledge territory. The third excludes the topic from knowledge altogether. An example is the advent of global warming. A prediction that it had 100 rather than 30 years to change the world's climate was a mistake. To say it is not happening at all suggests that it is irrelevant to consider seriously. A sector of the business world has treated the topic as a dangerous heresy, not to be mentioned in serious conversation, and thus lays the topic under a taboo. Until recent pressures to find the essence of issues, holism itself has been subjected to the same three modes of dismissal, as an error, as irrelevant and as placed under a taboo.

Each knowledge culture has developed effective boundary-riding techniques through allocating ignorance to all the others (figure 3.1). Individual knowledge is rejected by the others as biased; local knowledge as gossip; specialised knowledge as jargon; strategic knowledge as corrupt; and holistic knowledge as impractical. Familiar accusations are that specialist knowledge is too abstract to be useful. Community knowledge is seen as irrelevant since it can only refer to one case. Strategic knowledge of organisations is perceived as distorted because it seeks to influence reality, not merely to observe it. This rejection by each knowledge culture by the others forms a major impediment to synthesis.

For synthesis to succeed, the knowledge cultures need to be connected to one another sufficiently strongly to hear each other, and so to reach a collective understanding. This requires methods for establishing on-going relationships between them. These relationships need to be both synoptic (allowing the set of wholes to throw light on a shared theme in their own right) and synergistic (forming a set of interacting wholes that develop the theme further than before). Each knowledge culture in the decision-making set maintains its independence by building strong boundaries and isolating core practices. A synergistic approach requires that each sector allows materials from the other knowledge cultures to pass through its boundaries. Cores need to remain strong, but they also need to be made transparent to the other knowledge cores.

The boundaries can be considered as similar in structure to the semi-permeable membranes that surround each living cell. The membrane is strongly protective of the content of the cell. It is not inert, however, since it actively selects what will go in and out. The very success of the AIDS virus is due to its ability to trick the human cell's semi-permeable membrane and so pass through the cell's defences. For a synoptic outcome, knowledge and understanding (not just information) need to flow between knowledge cultures. The contributing knowledges can stay distinct, each with its own integrity. But the pattern they form is a holarchy, a system of wholes in which each is given full value, not a hierarchy in which some forms of knowledge are given more weight than others.

Where are we going?

Connecting the knowledge dots

So far, we have considered the set of Western knowledge cultures as a synoptic set, with each contributing separately to the Western decision-making system. Useful as this can be, it still does not tell us much about the potential for connections between them. In actual practice, the knowledge cultures are interacting with one another all the time. Their interactions are seldom made explicit, however, nor taken as seriously as they need to be. Moreover, the relationships between individuals, communities, specialists and organisations are only too often expected by all parties to be conflicts of interest. Two practical ways of making those connections explicit and collaborative are through linking the cultures in nested and in networked systems in which each knowledge culture is given equal weight. These can be represented from first principles fairly simply, as in figures 3.2 and 3.4, although in practice they form a dynamic, multi-layered system.

Figure 3.2 The knowledge cultures as a nested system (Brown 2005)

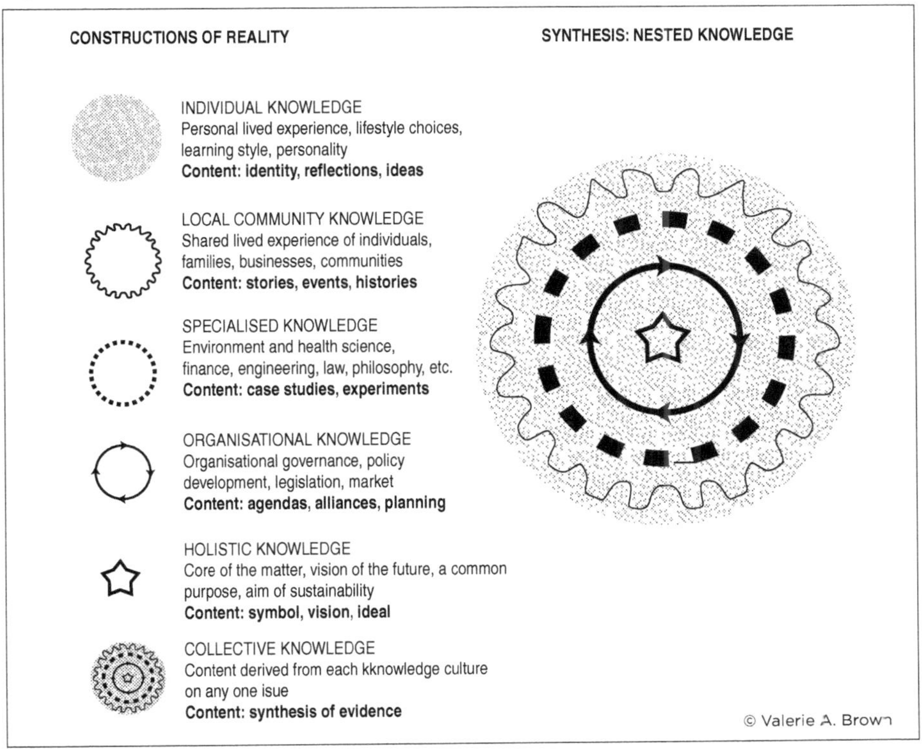

Collective decision-making can be considered as a sequence of nested knowledge cultures. The system begins, as does all knowledge, with the individual. In figure 3.2, individuals are represented as forming an almost invisible matrix from which all the other knowledge cultures are derived. The next layer is local community knowledge, formed from the shared symbols of place and events that unite all the individual experiences. This social basis for knowledge forms the context for the interpretations of reality by all the specialised disciplines, whether collected through interviews, observations, oral history, measurement or experiment. In the next knowledge layer, access to all the preceding knowledge cultures (individual, community and specialised) informs the strategic decisions of administrations and governments (figure 3.2). Finally, the knowledges can only be fitted together if they can arrive at some shared holistic focus, some agreement on a shared purpose.

As an example, consider the contributions to knowledge made by a cancer support group. Individual members have a unique and powerful experience, valid for each of them personally. In sharing their experiences, members of the group develop a deep community understanding of the experience of cancer, for themselves and for those working with them. The group may be asked to share this common understanding with the specialised researchers from medicine, sociology and psychology, trusting them to make good use of the information. The researcher in turn accepts an obligation to share their findings with the relevant organisations that have the capacity to act. Concern about a terminal disease is a powerful unifying factor, a holistic focus. This power flows back through the knowledge cultures. It eases the stress of the individual, gives strength to the support group, reinforces the value of the research, and is more likely to influence organisational change. The flow through the nested knowledge system is not only improved empirical knowledge of the disease, but also the human qualities of trust and compassion.

The ruling hierarchy among the knowledge cultures can distort the potential for synthesis offered by the nested pattern. For instance, the use of parallel social, economic and ecological accounting systems – known as the triple bottom line – has in practice increased competition for resources. The economic system, with its claims to a scientific basis, and its organisational pulling power, remains dominant. Social accounting systems are given lesser status as 'soft', and environmental accounts such as ecosystem services are still in their infancy. In contrast, an example of the effectiveness of collective decision-making is the response to the environmental pollutant, Lead.

There has been a long history of concern from all decision-making sectors about the health effects of the heavy metal Lead in the air. Researcher Deni Greene tells the story in *Reducing lead exposure in Australia: risk assessment and analysis of economic, social and environmental Impacts*, summarised in box 3.2. Individuals were first known to suffer

Box 3.2
DRIVERS OF CHANGE: PUTTING THE BRAKES ON HEALTH RISKS FROM LEAD

Only human industrial processes release the heavy metal Lead from the ground, so any health risks from the presence of Lead in the air are an inter-related people-environment issue. In 1935, the first diagnosed deaths from Lead heightened the level of community knowledge of the risk. Public health regulations against Lead in paint followed. As a result of continuing community concern, the Australian Academy of Science held its first inquiry into the social impact of Lead in 1965. By then, the metal was being released into the atmosphere from an exponentially increasing number of car exhausts.

Highly specialised biochemical assays determined the biophysical extent of the risk. Economists estimated the cost-benefit ratio of different permissible limits; clinical specialists examined the likely extent of the physiological damage. Specialised knowledge confirmed the risk, but offered no solution. The fragmented suggestions for possible control measures were rejected by government as too much of a burden on industry. Organisational knowledge held that the costs outweighed the risks.

Research between 1965 and 1985 found measurable health risks for adults above 10 micrograms of Lead per decilitre of blood. For children under six the risks are five times as great and include a one-point drop in intellectual capacity for every two extra micrograms of Lead in their blood. For adults, the lifestyle, and occupational options of those exposed to risk, such as smoking and working in a mine, increased the extent of the resulting physiological damage in adults. Yet the 'safe' level continued to stand at 25 micrograms per decilitre of blood, for both adults and children.

In the early 1990s the situation changed. Parents of lead-affected children found their voice. The Australian National Health and Medical Research Council found funding for a multidisciplinary community-based study of the social, economic and biophysical impact of Lead. The synoptic study by Berry, Garrard and Greene found that removing the Lead would cost tens of millions of dollars for industry adjustment. Leaving it in the air would cost hundreds of millions in health costs (1992 figures).

The synoptic combination of the specialist report, committed individual advocates and an awakening community conscience led to a change in the organisational construction of reality. After a series of round tables with all these interests present, the regulations were changed. Lead was removed from petrol over the following three years, and the permissible child Lead level reduced to 10 micrograms per decilitre of blood. Ten years later, the Lead in air levels in all the major Australian cities has dropped so low that monitoring has been discontinued. In Lead-smelting towns, however, there is quite another story.

The lesson seems to be that synoptic decision-making works; but for core social issues there is never a final answer.

Source: NH&MRC committee reports, 1989-95; Berry, Garrard and Greene 1992, Greene 2001

fatal neurological damage from lead paint exposure in the 1930s. By 1985, results of a 10-year study confirmed a permanent loss of intelligence in lead-affected children. It was 1995 before an alignment of parent and consumer activists, concerned communities and specialist research on the effect on children convinced a government advisory body to remove lead from petrol. The holistic catalyst was the united concern for the health of children. The lead story is an example of why we cannot afford to take account of some knowledge cultures and not others.

All five knowledge cultures, when taken together, carry decision-making into effective practice. However, the long-standing level of inequality between them means that positive remedial action is needed before you can have such equal input from all parties. We know what affirmative action means for gender and race: changes in language, resource allocation, ethical focus and social behaviour. What does it mean for synthesis?

In many non-Western cultures, for instance for the Australian Aboriginal people, collective thinking is the dominant knowledge form. Knowledge is developed for each community through elders' stories that weave together the individual, community, specialised and strategic understandings. Storytellers use 'song lines' that describe the landscape, the history of the community and the ethical rules as a single tale. Aboriginal and some Pacific island languages have no separate words for health or for environment; they are both expressed as the single word 'life'.

Western society methods of socialisation sub-divide that unity from the very beginning of learning. Divisions are established by the subject boundaries in schools, the distinction between what is learnt in school and at home, the limitations on the discussion of politics to avoid conflict, the segregation of communities into like-minded groups. It is not only too late, but counter-productive to wish to belittle the advances made within each knowledge culture. The challenge is to make the best use of them all.

In Leonardo da Vinci's work, he was meticulous in accessing all modes of thought. A contemporary writes that

Figure 3.3 Leonardo da Vinci. *Five grotescue heads*. Venice, Academia.

"when Leonardo wished to paint a figure he first considered what social standing and what nature it was to represent; whether noble or plebeian, gay or severe, troubled or serene, old or young, irate or quiet, good or evil; and when he had made up his mind, he went to places where he knew that people of that kind assembled and observed their faces, their manners, dresses and gestures ... After being satisfied with the material, he would proceed to give it shape". Leonardo's work included all the knowledge cultures and his drawings did not only include the socially acceptable. He recognised the significance of difference (figure 3.3).

How can we get there?

Networking the knowledge cultures

Placing the knowledge cultures in some ordered sequence, such as nesting the knowledges, is often required in research and in management. The emphasis is on developing the knowledge content, and the set can be entered at any point. In collective decision-making in the everyday world, the situation is likely to be much more dynamic and open-ended. The network of knowledge cultures in figure 3.4 provides some insights into such a fluid pattern. As with nested knowledge, the diagram in figure 3.4 is a simplified version built from first principles. In practice networked knowledge cultures interact with, and learn from, each other along multiple pathways. In a network, each knowledge process increases the effectiveness of the other; that is, they form a synergistic system. A working definition of networking is supplied by Donella Meadows (box 3.3).

Box 3.3

NETWORKS: CREATING A WEB OF CONNECTIONS

"A (human) network is non-hierarchical. It is a web of connections among equals. What holds it together is not force, obligation, material incentive or social contract, but rather shared values and the understanding that some tasks can be accomplished together that could never be accomplished separately. One of the important purposes of a network is simply to remind its members that they are not alone."

Source: Meadows, Meadows and Randers 1992

There is sound advice available for developing relationships between the different knowledge cultures. In 1971, Ivan Illich, the German sociologist of social change, published *Tools for Conviviality*. He pointed out that the festivals and celebrations of any society generate the synergy of whole-of-community thinking, releasing the collective capacity for imagination and creativity. His book argued that the industrial era, with its priority for objectivity, had masked the crucial role of conviviality in social wellbeing.

The idea of different forms of intelligence contributing to decision-making re-emerged in the 1980s. American psychologist Howard Gardner, in *Frames of mind,* identified the multiple intelligences that shape an individual's construction of knowledge.

Figure 3.4 The knowledge cultures as a networked system (Brown 2005)

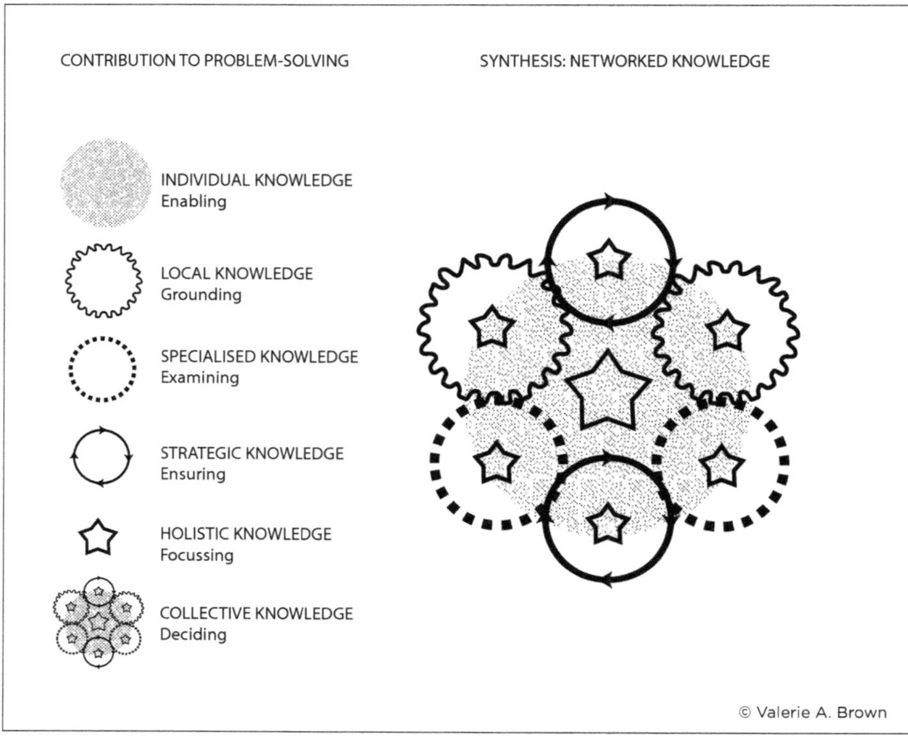

He accepted that there are emotional and aesthetic as well as rational forms of intelligence. All are needed in a lasting collective decision.

The French intellectual Michel Foucault explored the effects of the interactions that breach the strong controlling walls that separate the knowledge cultures. In dividing government from community, he suggests that the result has been the emergence of civil society, a negotiating space in which citizens act as a social conscience and limit the power of governments. This position is reviewed in some detail in the edited volume on *Power, truth, strategy*.

In the 1990s, American sociologist John Ralston Saul produced a strong critique of the 20th century belief that rational thinking can be detached from other ways of knowing. In *Voltaire's bastards: the dictatorship of reason in the West*, he points out that Voltaire's principles of democracy rested on the marriage of reason and ethical principles. If only reason is involved, Saul argued that any subsequent society is illegitimate.

Many efforts to break down the silo walls of the knowledge cultures have only served to increase the divisiveness. Inter-departmental committees established to forge organisational links can serve merely to protect the contributing territories. Community meetings monopolised by experts confirm perceptions about being marginalised rather than increase the community power to act. Anyone working to network the knowledge cultures needs to be aware of these risks. They can draw on the advice of Illich, St Boniface, Gardner, Foucault and Saul, to name only a few.

When taken together, each of the knowledge cultures makes a unique contribution to the pattern that connects. Individual knowledge *enables*, community knowledge *grounds*, specialised knowledge *examines*, organisational knowledge *ensures*, holistic knowledge *focuses* while collective knowledge *empowers*. All are required in any program of social change (figure 3.4). Networking the knowledge cultures generates synergies between them for the future, as well as enabling the immediate tasks.

Such synergy was demonstrated in the history of a social change strategy for the small coastal town of Shoalhaven in New South Wales, Australia. How a small group of myself, Jackie Ohlin, Jim Ireland and Rod Griffiths designed the project is described in *A great place to live with just a few things to fix*. The process of change was based on establishing networks between the decision-making knowledge cultures, in a manner that was explicit and transparent to the whole of the community concerned. The local relationships between the local knowledge cultures were assumed to be both synoptic (each knowledge culture complete in itself), and synergistic (knowledge cultures interacting to their mutual benefit).

The practical task before the change strategy team was to develop a structured plan incorporating a community-preferred future for the town. This future needed to be environmentally and socially sustainable in a location where pressures of change were making environment and community unsustainable. Shoalhaven was situated on the last undeveloped coastline within a day's journey of Sydney, one of the world's global cities. It was, therefore, exposed to extreme pressures of over-use of natural resources, uncontrolled urban development, and fluctuating employment.

To address this situation, an open learning process was created incorporating each of the decision-making cultures on equal terms. Translation and negotiating tools were crucial to the process, as was the decisions-into-practice framework that will be described in detail in the following chapter. The outcome is that long-lasting networks between the knowledge cultures have been established. A sub-district of Shoalhaven was able to achieve the following outcomes:

- Develop shared principles (what should be): they agreed on a development plan that met the interests of farmers, conservationists and estate agents;
- Describe the issues for people and place (what is): they argued constructively and successfully that a national highway scheduled to

be constructed through the centre of their town should become a bypass;

- Design for potential (what could be): they agreed on a structural plan that determined the long-term future of the area, with 12 formal objections to the plan instead of the usual 200 to 300; and
- Do in practice (what can be): they identified the indicators that would allow all parties (residents, experts, councillors and council staff) to monitor progress towards their agreed future, and collect the relevant data.

As with the nested knowledge cultures, the patterns formed by networking the decision-making sectors in figure 3.4, or in the case study above, are an outline only. The collage we are seeking in a social transformation towards collective thinking and action must be generated afresh in each case. We seek an adaptable framework, not a recipe. Contributions of the knowledge cultures may form any number of patterns, through forming and reforming fresh relationships according to the time and the context.

No-one was more aware of the connections between the knowledge cultures than Leonardo da Vinci. In his journals, he addressed the issues for achieving synthesis of different ways of thinking and working, ideally and practically. We have already quoted his ideal: "*Of what use is he who, in order to abridge the part of things of which he professes to give complete information, leaves out the greater part of the things of which the whole is composed?*".

And his pragmatic comment on why people do better work together than singly: "*It is far better to draw in company than alone for many reasons: the first is that you will be ashamed to be seen among the draftsmen if you are unskillful and this shame will cause you to study well. In the second place, a feeling of emulation will goad you to try to rank among others who are praised more than yourself. A third reason is that you will learn from the methods of such as are abler than you and if you are abler than the others you will profit by eschewing their faults and hearing yourself praised will increase your skill.*"

The rest of this journey towards synthesis will consist of finding the strategies and tools that could embed synthesis in the thinking and action of 21st century society. In the next chapter, we build on our understanding of the Western knowledge cultures to develop a synergistic decisions-into-practice framework. The framework brings the different knowledge processes together in an open learning system. We then describe some tools that have allowed the various knowledge cultures to make full use of the framework in pursuit of sustainable development. A prototype workshop is outlined in box 3.4.

Box 3.4

TESTING: CLASSIFYING THE DECISION-MAKERS (3 HOURS)

In this team exercise, ask participants at a multidisciplinary or community-based meeting to classify themselves as making their decisions based, respectively, on an individual, community, specialised, organisational or holistic perspective. Explain those terms using the diagram in figure 3.1. (15 minutes)

Place each group at their own round table. If there are not even numbers from each group, ask people to make up the numbers by adopting one of the positions. After all, everyone can think in each of those ways. Ask the whole group to agree on a shared priority issue. (15 minutes)

Ask each group to devise a collective solution to the issue, taking account of all five knowledge cultures. The solutions can be real or people can choose to explore theoretical ideas. (20 minutes)

Bring the groups back together to share their solutions for about 4 minutes each, summarised as 5 dot points. You can expect to be surprised at the differences between the groups' solutions. (20 minutes)

Then ask each group to share how they went about arriving at the solution. Hold a general discussion on these differences between the groups. Again, you will be surprised at the distinctive methods employed by the five groups. (20 minutes)

Invite the groups to each combine the collective strategies that they have designed to solve the priority issue they have all chosen. Each group to develop a collective action plan after learning more about the priorities of the others. (20 minutes)

Each group presents their plan to the plenary for critique and discussion, as a set of 10 dot points. (4 minutes each)

Discuss the differences remaining in the groups' solutions, and lessons for future action. (20 minutes)

In terms of outcomes, you would expect the workshop to have:

- aroused an understanding of the range of knowledge cultures involved in local issues;
- initiated a process for making collective decisions appropriate to local conditions; and
- generated sufficient synergy among the group to initiate a plan of practical collective action.

Note: This synoptic workshop process forms the basic unit for a wide range of social change initiatives, from a teaching/learning exercise, to the three-year research project described in chapter 7.

4 Collective decisions

"We travel not for trafficking alone;
By hotter winds our fiery hearts are
fanned:
For lust of knowing
what should not be known
We make the golden journey
to Samarkand"

James Elroy Flecker 1947 (1900)
The journey to Samarkand

A symphony orchestra and a guerilla army are both based on the principles of collective decision-making. Their members, the musicians and foot soldiers, act independently but remain commited to a`common cause. They build a strong community through each member contributing a particular skill. While maintaining their diversity and freedom of action, they nevertheless agree to follow a fixed set of rules. In their case, the rules are provided by a musical score and a tactical strategy. In collective action for whole-of-community change, individuals, community, specialists, organisations, and holists are bound by a similar need to maintain commitment, diversity and a shared set of rules.

What should be?

The decision-into-practice open learning spiral

The pathway to collective knowledge should not to be taken as a one-way track, along which the different knowledge cultures walk in single file. It is an ordered system of pathways that allow the travelers to learn from each other along the way, and then keep building on that learning. The goal is for the travelers to move together in an interactive, iterative process in which everyone enhances the understanding of everyone else. Like a spider's web, the different knowledge strands have considerably more strength when networked together.

Collective thinking does not call for starting afresh, but rather for extending the knowledge and skills we already have. Each of the examples in box 4.1 is based in a synthesis between a strong set of existing skills and open-ended potential, roots and wings. In Chapters 1, 2 and 3 we reviewed the divisions between the existing modes of decision-making and found them to be deeply divided between the knowledge cultures of Western society. From here on, rather than continue to look back on what to avoid, we will look forward to working towards *what should be*.

Box 4.1
COLLECTIVE SYNERGIES

In a synergy, the parts interact in a way that increases their effectiveness in the creation of a new whole*.

Examples are:

- Life-support cycles of Earth
- Renaissance cities
- Forests
- Google search engine
- Organisational executives
- Guerrilla warfare
- An orchestra

* Note: The value of the outcome rests on the principles driving the enterprise.

Again, we have a da Vinci example. Leonardo broke new ground in his powers of observation and level of inventiveness. He went even further, harnessing those skills to create a political statement, a collective understanding of *what should be*. *The Last Supper*, his painting of Jesus Christ and his 12 apostles dining together the night before his crucifixion, takes up one wall of a monastery dining refectory. The members of the group are painted as individuals in their own right, eating and drinking together in the refectory, just as their cloistered observers would be. This sent a powerful message of democracy to the monastery in a time of rigid church hierarchies. The relationships he captured between his subjects were so insightful that they have been the subject of many critical reviews. The most recent is the detailed and controversial analysis in Dan Brown's fictional *The da Vinci code*.

We cannot all match Leonardo's interpretive skills, and we live in different times. However, we do all have access to a shared mode of synthesis. David Kolb has described the learning process that underpins all adult constructions of knowledge. First reported in *Organizational psychology: an experiential approach* in 1974, the framework emerged from extensive studies of learning from social change among a wide range of occupations (figure 4.1).

Given the marked differences between the knowledge cultures, it is somewhat of a surprise to find that they all follow the same learning cycle. At one level, each culture's learning spiral is unique. Individuals learn through their own experience. Communities learn through their past events and expectations of the future. Science is built on objective inquiry. Organisational change is based on strategic planning and directed by the organisation's own agenda. Holistic knowledge is gained through

Figure 4.1 The individual decisions-into-practice open learning spiral d4-p4 (Adapted from Kolb 1974; Brown 2002)

(Note: the learning spiral is drawn as a cycle for ease of presentation)

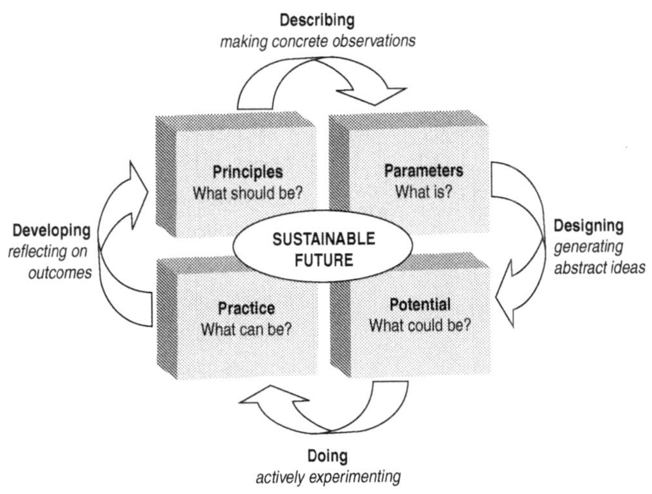

© Valerie A. Brown

a creative leap. Yet each follows the stages of the Kolb learning cycle, developing, describing, designing and doing (figure 4.1). For an individual, this cycle is the basis of reflective learning, for a community it is called community development. For a specialist, it is a research design, for an organisation a strategic plan. For an artist it is the creative leap. In times of change, the cycle becomes a spiral, each episode of learning building on the one before (figure 4.2).

Kolb and his colleagues found that the learning process begins with each person's set of guiding principles, their ideal of *what should be*. They check their ideal against what is, in order to evaluate what could actually be achieved. After this reality check, learners move on to explore the potential for new and novel ideas, to estimate *what could be*. After testing the new ideas in practice, they develop their own construction of a practical outcome, *what can be*. Learners then re-evaluate *what should be* in light of their learning experience. The learning cycle is, therefore, an unending spiral. It continues to be drawn as a cycle here, for ease of presentation.

For practical purposes, community educators such as Malcolm Knowles describe the learning stages (with the matching Kolb stages in brackets):

1. *Start where the learner is at* (developing primary principles, what should be);

2. *Undertake a reality check* (describing the parameters of the topic , what is);

3. *Introduce the new* (designing to use the imagination to optimise potential, what could be);

4. *Evaluate old and new together* (actively experimenting by doing, assessing what can be effective in practice): and

5. *Review where the learner has come to* (developing existing principles, what should be next).

and so on round the cycle again.

For each individual, the continuing spiral represents life-long learning. Kolb confirmed that only after the completion of a cycle, when learning has been tested against experience, does it become knowledge. Box 4.2 offers an opportunity to test this with your own learning from the experience of personal change.

Once Kolb and his colleagues had tested the cycle with enough groups of people, they found that, in any given occupation, one of the four stages is given greater emphasis. Administrators and organisations adapting to change emphasise the reflective *what should be* stage. Scientists remained focused on observations of *what is*. Successful managers of social change projects and holistic thinkers tended to make the imaginative leap of *what could be*. The professions remain concerned with the traditional empirical outcomes of their craft, with *what can be*. Thus, fragmentation between the knowledge cultures occurs between as well as within each of the learning stages.

The review of the Western construction of knowledge in chapter 3 makes it clear that conflicts of interest are only too likely to occur within each learning

Figure 4.2 The collective decisions-into-practice open learning spiral d4-p4*

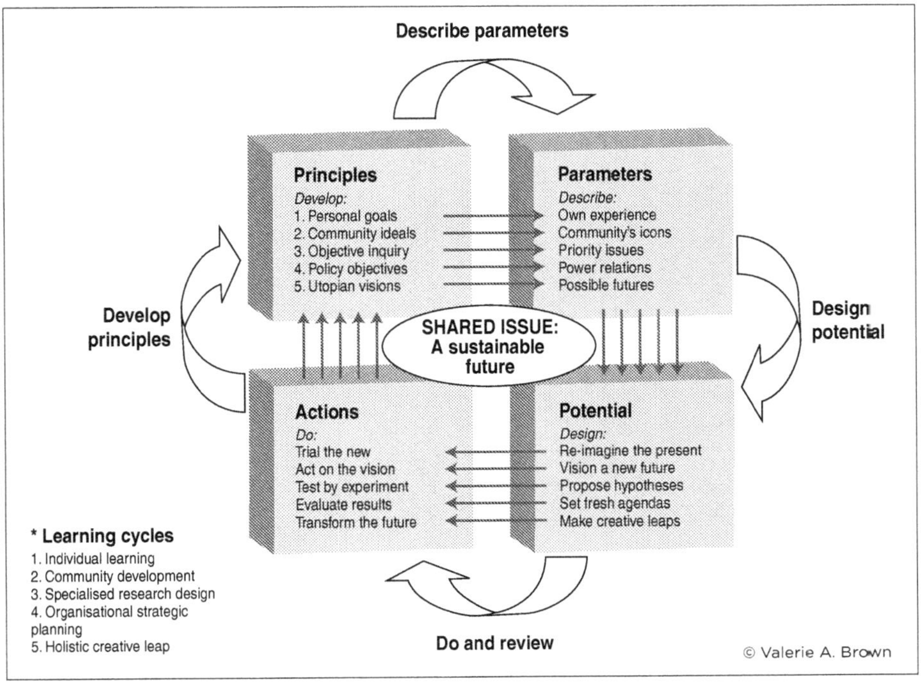

stage. The completion of the decisions-into-practice learning spiral will need tools to, first, ensure equal contribution from all the knowledge cultures at every stage, and second, to maximise the mutual learning within each of those stages. Leonardo da Vinci has been spectacularly successful in contributing to each of the knowledge cultures, but his amazing success lies in his use of them all together. There is no reason the rest of us cannot follow his lead. (see the several examples of collaborative information flows in box 4.3 below).

Memorable examples of collective social change highlight how whole-of-society learning may build up slowly, but the actual change can occur rapidly, as all the knowledge cultures come together. After President John F. Kennedy's assassination the world was never to seem as safe a place again. Changes in core symbols, such as the fall of the Berlin Wall between East and West Germany, or Galileo's telescope placing the sun not Earth at the centre of the universe, deliver a shared vision of a different reality. While the moment of change can be initiated from within any of the knowledge cultures, change follows from the combination of them all.

An example of an organisation initiating change is that of Shell Oil in 1995. After the global wave of protest at the proposal to sink the Brent Spar, a damaged oil platform, in the North Sea, Shell Oil redesigned their environmental policy, reversing many of their

> **Box 4.2**
> **TESTING CHANGE-ORIENTED LEARNING**
>
> Recall the last time you made a significant change-oriented decision. You will have started where you were at, reflecting on your purpose, and then checking on the important factors involved. Then you may have thought about potential outcomes and, if you could, found a way to evaluate the new idea in action. All of this may have passed in the blink of the eye, without your reflecting on your own thinking.
>
> If you do reflect, you will find that you have dealt with values (policies and principles), facts (observations and measurements about people and place), ideas (abstract thinking about potential change) and action (putting the ideas into practice), all in the one continual process. If any of the steps are left out, the decision-making process will have been incomplete, and you are likely to have to revisit the process to complete it later. This process applies whether you were making a decision about buying a new suit, or changing to a new career.

practices in using natural resources. A holistic contribution to a debate opens up immense possibilities for a new insight. Elvis Presley's adoption of the so-called black music of rock-and-roll is credited with accelerating cultural assimilation in the United States. The photograph of a small Vietnamese girl aflame from Napalm is considered to be the spark that helped end the Vietnam War.

Kolb's open learning cycle offers a synergistic framework for decision and action that is not a strait jacket. It follows plain commonsense. Each knowledge culture is free to expand its own reality, while working collectively towards some agreed purpose (figure 4.2). Each culture learns from the advances of the others. The collective decisions-into-practice spiral can be summed up in an acronym, d4-p4, shorthand for:

- **d**eveloping a common **p**urpose: answering the question *what should be*? Answers are expressed as principles, policies or value statements;
- **d**escribing the **p**arameters and paradoxes of the change in relation to that purpose: determining *what is*? Answers are supplied by the evidence collected by the different knowledge cultures;
- **d**esigning for **p**otential: discovering *what could be*? Answers are in the form of strategies created during dialogue between members of the knowledge cultures; and
- **d**oing in **p**ractice: testing *what can be*? Answers emerge from evaluation of the strategies as they are implemented collectively by the knowledge cultures.

The spiral continues as:

- **d**eveloping the **p**urpose further in the light of the new knowledge: *what should be next*?

And around the d4-p4 cycle yet again.

The decisions-into-practice spiral acts as a Rosetta Stone. The Rosetta Stone, which had the same message in Greek, Egyptian writing and Egyp-

Figure 4.3 Nested knowledge and the collective decisions-into-practice learning spiral (Brown 2006)

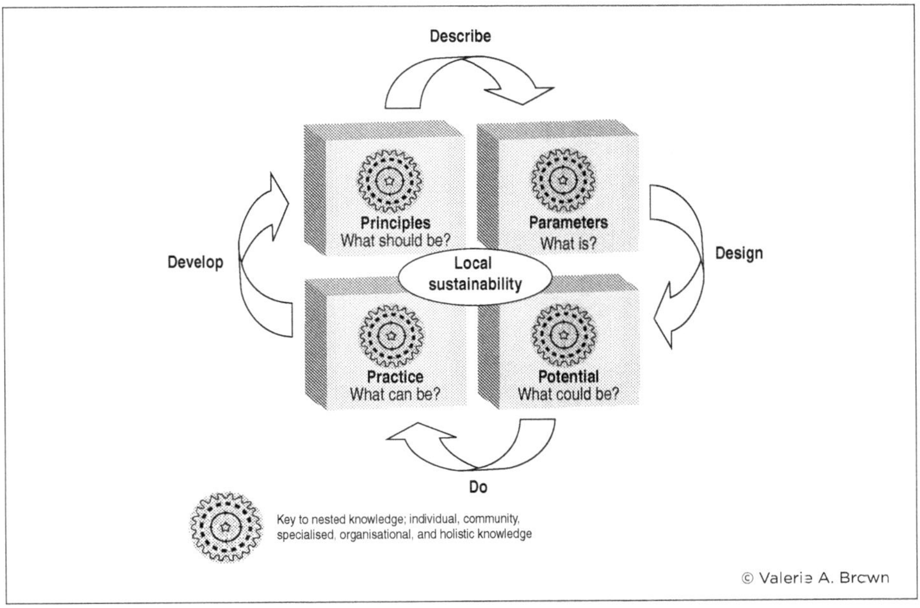

tian hieroglyphics written on its three sides, enabled the translation of long mysterious Egyptian hieroglyphics. In the same way, the d4-p4 spiral offers a framework common to each of the knowledge cultures so that each can understand the learning stages of any other (figure 4.2). The stages of the spiral are not set in stone, however. The spiral can be entered at any stage. As the wheel of social change turns, the stages of *what should be* (the principles), *what is* (the parameters of the change), *what could be* (the potential) and *what can be* (the practical) can change their order.

The next step is to identify the tools for synthesis of the knowledge cultures appropriate to each of the learning stages. These will involve combining the knowledge cultures in some mutually recognised pattern. As discussed in chapter 3. the pattern may be nested, networked or systemic. Figure 4.3 presents the learning spiral in relation to a nested knowledge pattern. Without such a pattern identifying a relationship between them, the knowledge cultures cannot make full use of the decisions-into-practice framework. Some of the synthesising tools are already well-used, some are in early stages of development, and some are being trialed as we write. A description of these tools makes up the rest of the book.

What is?

Closed, open and collective information systems

But before one can arrive at knowledge, there has to be information. Ways of accessing that information can be closed, open and collective. In Chapter 3 we learnt how the knowledge cultures maintain their differences. In the d4-p4 spiral we now have a framework of open learning capable of bringing them together. We now turn to tools that can enable synthesis at each of the mutual learning stages. The most basic of the tools are a common language and a shared information system.

A common language

The language of synthesis already exists, but has largely been neglected in favour of words for analysis. We have already felt the need to rescue words such as synergy, synoptic, and holarchy. The Oxford dictionary lists 156 words that use the prefix 'syn', referring to ways of being similar or joining together. Many of them, such as syncretic and syzergy, have, like 'sake' and 'quiddity', fallen into disuse. In thinking about the whole, they need to be revived. Re-orienting the language to accommodate changed ways of thinking is an honourable tradition. In the past half-century, the English language has changed to incorporate gender equity, racial harmony and peace. We have humankind rather than mankind; we talk about multi-culturalism, rather than mixed races. For those with a concern for a more peaceful world, 'target groups' and 'war chest' have been replaced by 'interested parties' and 'stored resources'.

A collective language will require the use of Standard English, rather than the phrases peculiar to any one of the knowledge cultures. There are, however, limitations when we decide to use English as the common language. Standard English has a built-in bias towards analysis and objectivity. For example, English has one word for 'language' and one for 'to know'. French has two words for each: *langue* for words and *parole* for meaning; *connaître* to recall facts and *savoir* to understand. Thus, in French, information and knowledge are clearly separated, while in English they can readily be confused. English has no equivalent of the German *zeitgeist*, the spirit of the times, or the Italian *la dolce vita*, living the good life. It is no wonder that the ideas of sustainability and sustainable development, focused as they are on a more just, peaceful and coordinated world, can be hard to explain in the English language.

Information systems

An open learning framework needs open, interactive information systems. As we found with the search engine Google in chapter 1, information technology can create information systems with the capacity for fresh synergies leading to new constructions of knowledge. Moving on from the vast academic libraries catalogued by the analytic Dewey system, three other types of information system have emerged. Open information systems generated by an organisation, a community group and a group of holistic computer professionals are described in box 4.3.

Each of the three systems in box 4.3 collects accredited information, but the accreditation process itself is markedly different. The Development Gateway is a hierarchical system, validated by experts from established contributing fields. Community-oriented designers put together the Open Knowledge Network based on a dynamic, open flow of information, validated by the users themselves. The communication space designed by the group Computer Professionals for the Public Sphere (CPPS) in *Pattern languages for civil communication* assumes that communication forms interactive patterns, evaluated by whether they add life to the society that generates them. The Wikipedia, an example of this last mode of information design, has been evaluated as slightly more accurate than the classic Encyclopedia Britannica.

The three information system designs illustrate the boundary conditions for

Box 4.3
LINES OF COMMUNICATION

Organisational closed system

In 1996, the then president of the World Bank, James Wolfensohn, announced that the institution had become a 'Knowledge Bank'. Broad consultation resulted in the design of an online system called the Development Gateway, a centralised database of specialised information, with a one-stop website, and the aim of combining 'usually inaccessible and fragmented knowledge'. The gateway is formed as a jigsaw of well-defined and pre-shaped pieces of information servicing the traditional knowledge framework, with high levels of reliability and accuracy. http://www.developmentgateway.org/EditorialPolicy [accessed January 2002]

Community-based open system

The Open Knowledge Network is based on a peer-to-peer filing system, generated by community online networks. There is no central database or command system. The system is highly flexible, with communication between the hubs reflecting the currency and urgency of issues at any one time. Associations of hubs, channels and contributors are based on mutual trust. The result is a shifting collage of community interests and messages. The process of creating knowledge is considered to be shared by all participants (Figure 4.2). http://dgroups.org/groups/OKN/docs/intellectualpropertyrights.doc [accessed September 2003]

Holistic system

Pattern Languages for Civic Communication is a project convened by the group Computer Professionals for the Public Sphere, based on Christopher Alexander's Pattern Languages. A pattern has a theme, an issue to resolve, a context and a proposed solution. Information stored in any one pattern connects in some way to any other pattern, just as words in a language find their meaning in relation to other words. In civic communication it is essential that all voices be heard. Information is regarded as seamless and placed in a live context so that knowledge is constructed collectively. http://www.cpsr.org/program/sphere/patterns/ [accessed September 2004]

collective knowledge making. There are no absolute divisions between information users and information providers in any of the systems. Permeable and blurred boundaries are common to all three. On the other hand, each type of design services its own constituency and maintains its power relationships. The global organisation keeps control of its information, preserving its dominant position in the hierarchy. The community-based system builds its information system on the presence of mutual trust, and maximises open learning. The holistic pattern language seeks to establish cognitive justice, equalising the contributions of all the knowledge cultures.

In our tightly compartmentalised society, there are regularly calls to abolish boundaries; to wish for the freedom of 'boundary-lessness'. This is an idle wish. There is no such thing as boundary-less knowledge. It is the borders between knowledge and ignorance that have created the strength of each knowledge territory. The boundaries provide the protection necessary for the cores to develop in the first place. Recognising that there is a knowledge culture behind each information system is a basic step when gathering information to support each learning stage.

Where could we be?

Strategies for synthesis and social change

The decision-making open learning spiral can be entered at any stage of the learning. When the goal is to guide collective social change, however, it is crucial to start with the first stage, in order to establish the set of shared principles that will direct the learning process. These principles need not be the same for each knowledge culture, but they must be compatible. The Earth Charter, described in chapter 8, is an example of a coherent set of principles generated by all the knowledge cultures.

Principles are necessary guides to social change. Where the goal is to continue in the same direction, more of the same, the principles tend to be taken for granted. The learning process ignores the first stage and begin with describing what is. The father of social policy, Richard Titmuss, in his classic *The problems of social policy,* suggests that formulating a policy in itself signals an intention to change. Policy development is not needed if the intention is to continue as before.

The choice of tools to structure each learning stage is critical to the collective process. Shared principles can readily deteriorate into meaningless 'do good' phrases. Different sources of information can lead to endless 'Yes it is', 'No it isn't' debates. Maximising the potential contribution of each knowledge culture as separate exercises can increase rather than bridge potential conflicts of interest. Setting out to test the collective learning in separate teams can sabotage the entire process. Strategic tools are needed to give full expression to each knowledge culture's deeply felt goals, their different sources of information, their creative capacity and their practical skills. The criteria for suggesting the tools in box 4.4 are that they work in practice, they are convivial to use, and they promote equity among the knowledge cultures.

Box 4.4

TOOLS FOR SYNTHESIS AT EACH LEARNING STAGE: IDEALS, FACTS, IDEAS AND ACTIONS

Stage 1. Ideals: combining different ideals as shared principles asks for mapping personal constructs among the knowledge cultures, opening up a mutual understanding of each other's purpose.

Stage 2. Facts: combining different sets of facts involves interpreting the discourses in which the evidence has been developed.

Stage 3. Ideas: combining different creative ideas calls for following the rules of dialogue so as to enable groups to celebrate their differences.

Stage 4. Actions: combining different actions involves approaches to resolving wicked problems which accept ambiguity and paradox.

Developing principles: Construct mapping of alternative futures

The American engineer-turned-psychologist George Kelly in *The psychology of personal constructs* offers a method for reconciling different value positions. Kelly differs from many other psychologists in assuming that each human being is also a psychologist in their own right. It follows that each individual engaged in a collective exercise is capable of evaluating the connections between their position and that of others. This approach avoids the loggerheads of fixed, non-negotiable positions. Kelly's 1960s research found that, in order to make sense of

Figure 4.4 Alternative futures (after Kelly 1965)

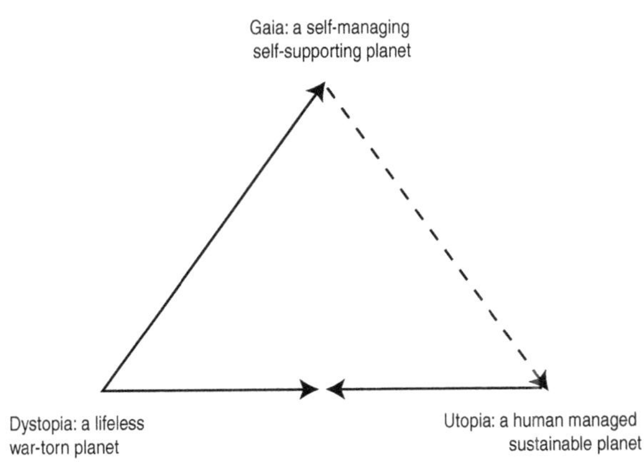

a situation, we position ourselves along a continuum between two poles. Our position between the two extremes determines how we construct reality. The introduction of a third or more possible poles opens up a dialogue that allows each individual to understand the others. It also allows a collective set of principles which encompass all the poles, rather than maintain the oppositions.

The classic example is the twin poles of love and hate. If there are changes to be made in a relationship, that one dimension offers only a forced choice. The many other possible opposites to love include friendship, detachment, apathy and objectivity. Opposites to hate include righteous anger, neutrality, and tolerance. Opening oneself to any of these expands the available options for a new relationship. Kelly's work is used in counselling to allow distressed participants to consider possible alternatives to existing relationships, and so open up avenues for change. It can also be used to inform the direction of social change.

Following Kelly's method of personal construct mapping, figure 4.4 presents alternative global futures. A prevailing idea is that the world is locked in a choice between unspeakable disaster and an improbable utopia (the baseline of the triangle in figure 4.4). Identifying a third possibility turns the single forced-choice dimension into a creative decision-making triangle. The options for the future change from a zero-sum game (if your agenda wins, mine loses) to a positive sum game (an enhanced outcome for the entire enterprise).

Construct mapping is useful as a tool for diagnosing differences within a community, among specialists, and between organisations, without attributing praise or blame. It is equally of value in promoting alliances between them. Box 4.5 describes a workshop in which the tool was used as the basis for a collective understanding between workshop organisers. The same exercise, combining six poles, is applied to integrative inquiry in chapter 7.

Describing the parameters: interpreting the discourses

Evidence is assumed to be based on fact. Facts are conclusions about different aspects of reality, based on specified sources of evidence. So there are different bases for determining fact. Since each knowledge culture draws on a different source of evidence, their facts will have different foundations and will need to be reconciled. Dutch sociologist Martin Hajer uses the idea of a discourse as a route by which different knowledge platforms can be understood. He applies the term discourse as in Descartes' classic *Discourse on the method of rightly conducting the reason*. Hajer distinguishes discourse (a set of connected ideas) from discussion (unstructured communication) and debate (opposing communications).

Hajer undertook a series of case studies of local communities working towards a more ecologically sustainable future. He identified four discourses on the management of the change, all contained within the dominant social discourse of the late 20th century. The dominant discourse he clumsily but accurately entitled 'ecological modernisation', referring to the use of modern technology to resolve issues arising between social and natural environments. In a synthesis of the evidence, there is need to reconcile the claims made by the various socio-political discourses. Hajer found that these are, in a continuum from technology-driven to socially driven (with Hajer's terms in brackets):

- Society changes the environment through the use of technology (*rationalisation of ecology*);
- Society responds to the needs of environmental systems through the use of technology (*technicalisation of ecology*);
- Society changes itself through applying technologies sympathetic to environmental systems (*ecologicalisation of the social*); and
- Society changes itself by applying the rules of ecology to social systems (*socialisation of ecology*).

Hajer's work provides an example of the range of possible positions in one arena, the management of social and environmental change. It also provides an example of how valuable ideas from one knowledge culture - here, the specialist - can be incomprehensible to the others through the use of specialised language. Each knowledge culture involved in any decision-making on the future will have their own set of dominant discourses, just as they have their own preferred information system. How to bring these together in and between each knowledge culture is explored in chapters 5-9.

Box 4.5

A WORKSHOP ON SOCIAL LEARNING PROCESSES

A group of colleagues designing a workshop shared their aims for the workshop through Kelly's personal construct theory. Each pair of poles delineates a concept or idea. Any one interpretation can have many opposites, each pair making up a different underlying concept.

In the concept map, the upper layer identifies the organisers' principles for designing the learning process, the lower layer the principles for the participant's experience. Note the marked difference in workshop design principles, depending on which poles are chosen. The challenge to the workshop organisers was to include them all.

Source: Brown et al in Keen, Brown and Dyball 2005

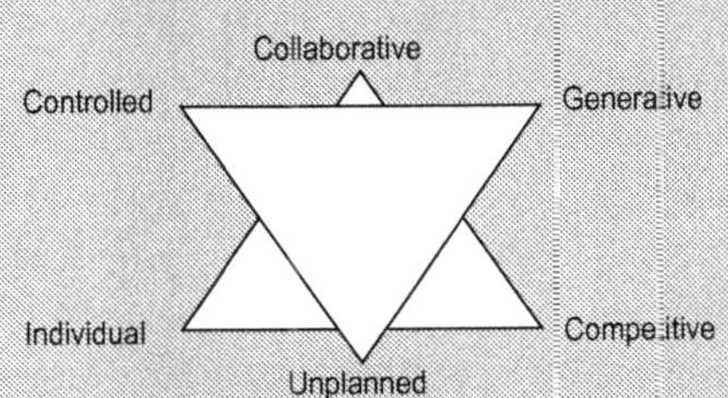

© Valerie A. Brown

Designing what could be: Following the rules of dialogue

In human communication the actual words (*parole*) carry subtle underlying meanings within the cultural discourse (*langue*). Physicist-turned-epistemologist David Bohm argues that mutual dialogue is essential for any effective cultural interaction. The way people learn, he says, is through a celebration of difference - not through opposition and debate, and not from hearing what they expect to hear. The derivation of the word dialogue is not, as is often assumed, talk between two. Its Latin root means 'through speaking'. Bohmian dialogue does not mean just any interchange, but following a set of rules of dialogue. Those in box 4.6 were derived from Bohm's work by Phillip and Marsha Snow Gang, Montessori-based educators. They drew on their practical experience in applying Bohm's ideas in an online post-graduate degree in integrated studies. While the rules are familiar to trained facilitators, they require considerable self-discipline and guidance when they are adopted, as Bohm intended, by everyone involved in a change process.

Box 4.6

RULES OF EFFECTIVE DIALOGUE

Throughout the dialogue:

1. Commit yourself to the process.
2. Listen and speak without judgment.
3. Identify your own and others' assumptions of reality.
4. Respect other speakers and value their opinions.
5 Balance inquiry and advocacy.
6. Relax your need for any particular outcome.
7. Listen to yourself and speak when moved to.
8. Take it easy – go with the flow – enjoy.

Source: Drawn from Bohm 1996 by Gang, P. and M., *The Institute for Educational Studies* (TIES) 2004

The rules in box 4.6 differ markedly from the standard pattern of a conversation in which one puts one's own case. We typically reply to an opinion from others, with one of our own: "My own opinion is ...'. In contrast, the first step in dialogue is to put your own case aside, and commit yourself to open-minded learning from the others. 'Please tell me more about your idea...'

The second step, listening and speaking without making a judgment, is more difficult than it may seem. We readily make the reply 'I think you are right/ wrong...'. More can be learnt from listening to positions with which you disagree than from ones with which you agree. A dialogue-type response would be 'I haven't heard that before.........can you explain further?'

Critical questioning is common to Socratic dialogue, scientific method and Bohmian dialogue alike, but the purpose differs in each case. The Socratic approach maintains an atmosphere of critical doubt. Scientific method means abandoning one's own perspective in order to make objective observations. In dialogue, participants question each other's assumptions in the spirit of critical loyalty described by Nigel Dower. Kelly's construct mapping is useful here. 'Do you mean that..........?'

The fourth step, respecting each other's ideas, does not mean either agreeing with or correcting them. It is about indicating respect for what they have told you, and checking that you have heard them correctly: 'Do I understand you correctly ?'. Fifthly, inquiry and advocacy are easily confused. Each participant needs to question their own assumptions in asking questions and exploring difference. Six, remaining free of a predetermined outcome leaves the conversation open to new and innovative ideas.

As the American psychologist Howard Gardner has demonstrated in *Frames of Mind*, there are many ways of using your intelligence, of which rationality is only one. There are also emotional and intuitive forms of intelligence. In his seventh point, Bohm suggests that you listen to those other voices and introduce them into the discussion whenever you feel it is appropriate.

The eighth rule refers to the experience of taking part in a dialogue. The differences in the interactions should generate conviviality and interest, not argument. This brings to the activity the exhilaration of mutual problem solving. Examples of this creative leap will be discussed for each knowledge culture in chapters 5 to 9. Figure 4.5 identifies the differences between a course based on the prevailing Western rules of learning, and one in the spirit of dialogue-based learning.

Doing in practice: resolving wicked problems

In working towards a collective ideal, we have no choice but to start from wherever we are now. No matter how innovative the potential new directions, the learning cycle foretells that if we are to establish new knowledge we must eventually come to 'doing and reviewing the practice'; that is, testing the new in the context of the old. With wicked problems where there are complex, interrelated issues, there can appear to be no place to start. The issues are an inherent part of the society that seeks to resolve them. The key characteristics of wicked problems are explored below. Such problems cannot be resolved by employing solely the analytic mode of thinking in which the issues are simplified to a few key factors or referred to existing solutions.

Wicked problems concealed within most communities today include violence, homelessness, climate change, global warming, recreational drug use and the ubiquitous presence of silent chemical pollutants. Issues such as these are whole-of-society problems. When they are

Figure 4.5 Traditional and dialogue-based learning

Component	Traditional	<<...........>> Dialogue-based
Learning unit	Bounded, one-off	Open, cumulative (d4-p4)
Knowledge content	Fixed, expert evaluated	Emergent, group generated
Knowledge source	Leaders in the field	All participants
Learning channels	Teacher to learner	Co-learning
Learning style	Analysis and debate	Synthesis and dialogue
Knowledge reliability	Empirical data	5 types of evidence (chapter 7)
Communication process	Objective criticism	Critical loyalty

endemic in a society, it is not possible to eliminate the society as one would eliminate a risk. Following Kelly, each of the society's problems is almost certainly another face of that society's areas of strength. Wicked problems are not solved; they are part of the challenge of guided social change. According to Rittel and his team, the following forms of paradox typify the wicked problems that accompany social change:

1. Wicked problems are those without any final solution, but nevertheless urgently requiring an answer.
2. Every wicked problem poses a wide diversity of questions, but requires a unified response.
3. A wicked problem requires an unprecedented solution, but has to be considered in the light of existing solutions.
4. Wicked problems require the reconciliation of the two logically distinct yardsticks of facts and values.
5. Wicked problems contain an imperative to change, but offer no assurance on which directions to choose.

Describing the task of addressing wicked problems, not as 'solving currently unsolvable problems' but as 'resolving paradoxes', gives entry to pathways for collective social change. The example in box 4.7 is of global warming, a key issue for a sustainable future, expressed as a set of paradoxes. Each paradox becomes a useful

Figure 4.6 Resolving wicked problems

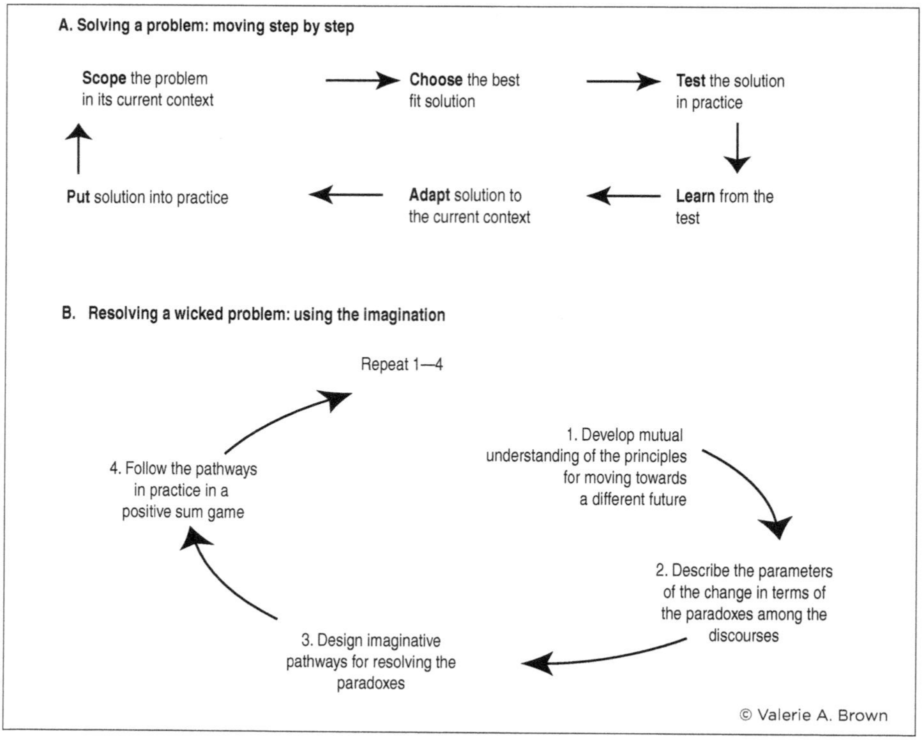

fulcrum for change. The change has no predetermined course; it will be generated by the interaction between all the interest groups concerned. There is no given correct solution; the solutions generated in the decisions-into-practice process cannot be known beforehand. Concept mapping, identifying discourses and taking part in dialogue are all valuable tools for resolving paradox.

Accepting the ambiguity and uncertainty presented in addressing wicked problems, and taking a positive approach to paradox rather than

Box 4.7

GLOBAL WARMING ... A WICKED PROBLEM

Global warming refers to the raising of the global temperature through carbon dioxide and other heat-trapping gases in the atmosphere. The gases are produced by the fossil fuels used to provide energy for industrial and domestic use. The temperature rise is accompanied by changes to rainfall and storm intensity. Consequently, agricultural conditions are changing, natural disasters are increasing, and so is the spread of disease. Collectively, the greenhouse gases threaten critical global life support systems.

Issue Type 1. There is no final solution

Global warming displays three paradoxes. The source of the problem, fossil fuels, has at the same time been the foundation of great progress in human well being. Equity between countries relies on developing countries increasing their energy use, but this increases their greenhouse gases. Solutions using wind, water and sun are known, but are not considered viable in dominant economic theories, since these do not take account of the damage from excessive fossil fuel use.

Issue Type 2. Every problem is unique

Periodic global warming and cooling are predictable cyclical events over geological time, but the current change is unprecedented in both speed and cause, as the by-product of a single species in only a few centuries.

Issue Type 3. Limited by existing solutions

Solutions in the post-industrial era still look to technology. Global warming arises from human actions.

Issue Type 4. Combining facts and values

The causes of global warming and its solutions are technically well understood. The potential solutions that can fit the cultural values of the contributing countries are as yet unknown.

Issue Type 5: No solution without further problems

Global warming has no currently imaginable solution for the well being of the human species other than restoring the global balance by returning to the atmospheric mix of before the industrial era. This poses new problems from the impact of new technology, or the reduction in the use of present technology, or both.

Source of material on global warming: McMichael A.J., 2001, *Human Frontiers, Environments and Disease*

Figure 4.7 Local Sustainability Project field studies, 1996–2003

Decision-making base	Focus of field study	Outcome of field study*
Individual	Transdisciplinary seminars Individual inquiry	PhD seminar program PhD theses (chapter 5)
Community	Regional collaboration Regional monitoring	Whole-of-community toolkit Regional State-of-the Environment Report (chapter 6)
Specialisation	Professional practice Professional education	Social change action web Sustainability and health text (chapter 7)
Organisation	Whole-of-council change Local government advocacy	Integrated triple bottom line Baton Forum for local sustainability (chapter 8)
Holist	Pattern making A pattern language	Holistic practice Liberating voices: living communication for the public sphere (chapter 9)
Collective	Synoptic workshops	Integrating learning from each of the above (chapter 10)

*Published as UWS PhD theses; Aslin and Brown 2004; Brown et al 2000; Nicholson, Stephenson, Brown and Mitchell 2002; Brown, Grootjans, Ritchie, Townsend and Verrinder 2005; Keen, Brown and Dyball 2005; Paine, 2005; and Schuler 2006, respectively. See references for details.

trying to eliminate it, is no easy matter in a culture that has long preferred the simplest possible solution. The more complex approaches involved in resolving wicked problems requires some practice. Figure 4.6 contains the steps followed in usual problem-solving practice, and in collective problem-solving. The exercise in box 4.8 at the end of the chapter provides an opportunity for that practice.

Where can we be?

Design of a research study

We have now developed a framework for bringing the knowledge cultures together in an open-ended learning spiral, d4-p4. Following the framework involves taking four steps: sharing the values of the different knowledge cultures through concept mapping; pooling their evidence by identifying their various discourses; generating creative solutions through the use of dialogue; and acting on the ideas and the evidence through accepting the notion that these are wicked problems that involve paradox. We have thereby suggested a suite of tools for harmoniously bringing together the ideals, sources of evidence, creative ideas and practical actions in collective thinking and action.

The next question must be: how does all this work in practice? The framework and tools offered here have been tested in a collaborative action research program. Between 1996 and 2003, the Local Sustainability Project, of which I was the coordinator, undertook a study of collective social change originating in each of the knowledge cultures in turn (figure 4.7).

The results of the studies provided the insights into the principles, parameters, potential and practice of the collective management of social change reported in chapters 5 to 9. There were 10 field studies of sustainability as social transformation projects, two from each of the five decision-making cultures. In each study, a sixth knowledge culture was generated by the project itself. A collective decision-making process of synoptic workshops brought together members of each of the knowledge cultures on equal terms. The characteristics of that collective knowledge process are reviewed in chapter 10.

The study was conducted in the spirit of practicing what you preach; that is, in terms of applying collective approaches to the research process itself. Such a practical purpose for an idealistic enterprise is well expressed in Plato's comment on trying to implement Socratic ideals in his utopian design for The Republic:

"On seeing beautiful creatures, whether works of art, or actually live but in repose, a man should be moved to desire to behold them in motion and vigorously engaged in some exercise as seemed suitable to their physique. Well, that is the very feeling I have regarding the descriptions of the ideal state."

Plato, on Socrates in *The Republic*.

The various projects that make up the Local Sustainability Study, taken together, offer the opportunity to achieve Plato's goal of seeing the ideal come alive. From this point on, we make use of the social learning framework, open communication system, wicked problem resolution, and collaborative language that have been developed so far. We move on to examine how each knowledge culture puts together the working parts of a collective decision in working with social change. The next chapter addresses the role of the individual in generating collective social change. Before you continue, consider working through the exercises in boxes 4.8 and 4.9.

Box 4.8

TESTING: RESOLVING WICKED PROBLEMS

Choose for yourself any wicked problem in your own community that touches you personally and seems unsolvable under present social conditions: for example, illegal drug taking, domestic violence, global warming, and the increasing gulf between rich and poor. Consider you are currently addressing that problem. Use figure 4.6 above to identify the differences between the usual process, and the following:

1. Diagnose the purpose, what should be, in resolving the wicked problem, by personal construct mapping involving all the knowledge cultures;

2. Describe the parameters of the issue through searching on Google, identifying the knowledge discourses that are linked to the issue, then compiling the evidence from all the knowledge cultures;

3. Design a potential solution through holding a dialogue on innovative solutions with three other people who are also interested in the issue, if possible, each from a different knowledge culture;

4. Do in practice the actions that implement the selected solution, making sure to involve each of the knowledge cultures, and to start reflecting on events from the beginning; and

5. Reflect on your own learning from the exercise, and be prepared to go round the cycle again - and again.

Box 4.9

TRANSLATING THE COLLECTIVE LEARNING SPIRAL INTO PRACTICE*

*NB This workshop does not stand alone. It provides the bridge between 6-12 months of collective skills development (described in chapters 5-9 of *Leonardo's vision*); and the collective implementation of the resultant action plan (chapter 10).

Aim: to establish whole-of-community collective decisions and action on a complex issue. Community: a group of people bonded by a shared commitment to place (local, regional, national, or global) or strong and continuing mutual interests (personal, professional, organisational or cultural).

Participants: Chairperson, administrator, facilitator, recorder and observer/integrator and 10 - 100 workshop participants (roles may overlap). Workshop numbers should be evenly balanced between key individuals committed to resolving the issue, local citizens concerned about the issue, specialists with expertise on the issue, representatives of influential organisations (including government), and holists (integrative thinkers). Each working group has 2 people taking up the interests of each of these 5 groups.

Resources: A pleasant, neutral site, separate from any of the interest groups. A round table for each set of ten participants. Each table has 4 large sheets of stiff drawing paper and 4 A3 sheets, each marked with one of the 4 questions in Figure 1. Coloured pens, bue tac, water, refreshments, a data projector, screen and whiteboard. Arrangements for recording and reporting back to participants within 1-2 weeks.

Box 4.9 continued

Figure 1. Collective decisions-into-practice social learning spiral (d4-p4: Brown 2001)

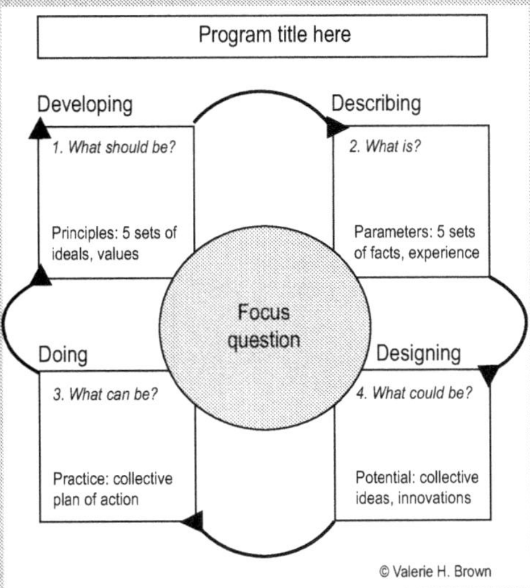

Timetable: Time: from 3 hours to 3 days. Step 1. First 20 minutes in plenary discussion of the collective learning spiral as a guide to collective change. Step 6. The final 20 minutes is spent in plenary adopting the action outcomes of Stage 4. The rest of the time available is divided between the 4 questions (Steps 2-5). For each question, each group member writes their answers to the commonsense question on each large sheet for themselves, in 5 words or less. Answers to the question are reviewed within the group in a constructive (not critical) dialogue. No answer is right or wrong. After discussion, the answers may be altered or combined, but only the person who wrote it can change or transfer a response. No-one alters another's item on the sheet. In the last 10 minutes of the time allotted for each question, the answers are collated onto the A3 sheet which is then placed on the wall for inspection by the whole group.

Synergy: Be upbeat through this whole process. The outcome is not intended to be consensus or the lowest common denominator. It is a rich picture of what can happen in this place. Anything can be achieved if all interests can support the outcome. Negotiation, dialogue and trust are the essence of the workshop. With 5 groups or less, the participants can develop a joint action plan. Over 5 groups, the workshop team puts a synthesised action plan to the whole group for their collective commitment to the outcome. Follow-up tasks (meeting schedules, selecting and monitoring indicators of progress, liaison arrangements, information sharing) need to be allocated to 'volunteers' from each project group.

5 Thinking for oneself

"I celebrate myself,

And what I assume, you shall assume,

For every atom belonging to me as good as belongs to you

I am large.....I contain multitudes...

Failing to fetch me at first keep encouraged,

Missing me in one place search another,

I stop somewhere waiting for you."

Walt Whitman 1996 (1855)
Song of myself (opening and closing lines)

Never ask whether an individual can make a difference. They can, but not alone. Adolf Hitler, Mother Teresa, and Bob Geldof are only a few of the multitude of individuals who have been catalysts for whole-of-society change. Whatever the reason anyone seeks to transform society, they need the capacity to bring others along with them. One may wish to relentlessly impose their own vision on a docile population (think Hitler). Another may wish to recruit others to pursue a common cause (think Mother Teresa and Geldof). There is a third choice, not either/or, but a willingness to do both, to work in harmony with others and leap towards one's own vision at the same time (think Leonardo da Vinci, Einstein, and the designers of innovative structures such as Google).

What should be:
the capacity for collective genius

Every human being is born with an age-old brain and a brand new mind. The biological program that has created the physical capacity to think is at least 2 million years old. Work by Jean Piaget and his successors has shown that the capacity to see, hear and learn is constructed afresh on their biological base with the birth of every human being. At the same time, there are generic human capacities that emerge in every generation and all times and places. One of these is the power of social communication. Another is the power of the mind that we have come to call genius.

Approaching the two synoptically, they can greatly enhance one another. Given the right conditions, the social and the personal interact to form a creative sum game, one in which the result is more than the contributions. Of course, the Hitler example is proof that this game can spiral out of control. The examples in box 5.1 are a few of the many individuals who have played this double role, maximising their own individuality and contributing to a common interest at the same time – for better or for worse.

Box 5.1

WITH A LITTLE HELP FROM MY FRIENDS: INDIVIDUAL LEADERSHIP IN A COLLECTIVE UNDERSTANDING

As one of a team:

- The Beatles
- Orchestra conductor
- Theatrical producer
- A sporting team
- Mother Teresa
- Paulo Friere
- Bob Geldof

As the symbol for a whole:

- Leonardo da Vinci - Renaissance
- Dalai Lama - Religious tolerance
- Babe Ruth - Sport
- Marilyn Monroe - Hollywood
- Albert Einstein - Science
- Milton Friedman - Market economics
- Adolf Hitler - Fascism

In this chapter we will find that human thinking has always been collective and served both individual and the society. We identify the conditions that could lead to genius, and discuss how this applies to collective thinking. Building on these generic human abilities, we first consider how each individual can access their own collective capacities, and then how they can build upon the collective capacity of others.

Two recent Renaissance thinkers can help us again. Anthropologist Clifford Geertz in *Local Knowledge* concludes that the basis for individual decisions is 'common sense'. Common sense, he suggests, is another name for the synergy that springs from each person's experience as an individual, a community member and a subject of organisations, simultaneously. Mostly this synthesis is instinctive, completed without conscious effort. Biologist-philosopher Michael Polanyi in *Personal Knowledge* points out that the major part of our individual knowledge is tacit; that is, assumed, and rarely examined.

Under conditions of social change, tacit knowledge is called upon in unexpected ways. Using one's own common sense becomes part of a learning process linked to other people's common sense. Accessing one's own commonsense is not always straightforward. The educationist, Paulo Friere, as a social activist in Brazil, gave members of a displaced, despairing community instant cameras to photograph what they valued in their community. In bringing this tacit knowledge into consciousness, members of that hopeless community gained a sense of identity and hope for the future. In this case, enabling an individual's reflectivity (that is, learning from reflection on one's own learning) from looking at the photographs, resulted in a sense of purpose and direction – and in taking action to improve their situation. The process of reflectivity allows each individual to identify their own and other's individual approach to change (figure 5.1).

Most of us have a personal preference lying somewhere between 'wanting to change everything all the time', and 'never willing to change anything at all'. We might be committed to the outcome of a change, but be unsympathetic to the process, or vice versa. Some things we may want to change, some we don't. Treating change as a creative sum game requires that all the positions on the change continuum have been considered. They can be related to each other, not defensively or dismissively, but synergistically. For the purposes of this chapter, it is assumed that we are referring to the individual who holds the 'initiating change' position on the continuum.

From this position, guiding change through all the stages of the open learning spiral, as described in the previous chapter, may appear too complex for any individual. Fortunately, there is help. Our brains have a most powerful unifying capacity. Nobel Laureate in Neuropsychology, Alexander Luria, records how accident-damaged brains – even if 50% of the brain is destroyed – can restore

Figure 5.1 Options for working with change

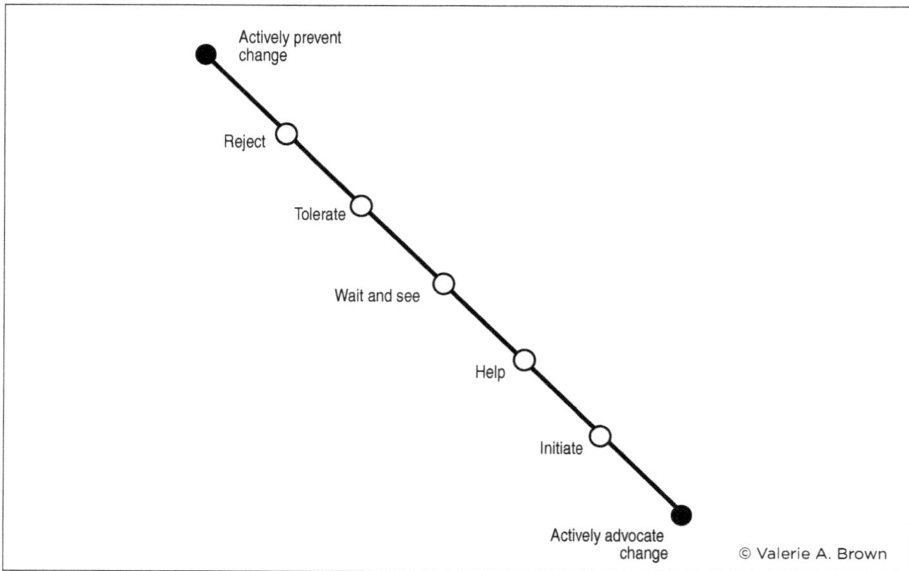

the identity of the person, and reunite previous learning, if the frontal lobes and optic chiasma remain. If anyone loses the use of their hands, and learns to sign with their foot, the signature is still legally recognisable. These are strong supporting conditions for the individual capacity to collect their own thinking into a single powerful synthesis. The primary support is the access to the thinking of all those that have gone before.

Scholars have always been aware that knowledge is collectively created over time. To quote Bernard of Chartres, as long ago as the 12th century: "*We are like dwarfs on the shoulders of giants, so that we can see more than they, and things at a greater distance, not by virtue of any greater sharpness of sight on our part, or any physical distinction, but because we are carried high and raised up by their giant size.*" Since that time, the same statement has been adopted by Isaac Newton, on the rim of a British coin, and as the motto of the reference system Google scholar.

Consider Leonardo da Vinci with his machines and his paintings, van Leeuwenhoek's discovery of the cell as the unit of life, and Einstein's of the relativity of time and space. In their own time each seemed to their contemporaries to mark the limits of what humans could know. Leonardo's master stopped painting himself when he saw Leonardo's first painting of an angel. Yet each of these examples built on the ideas of their predecessors and provided the springboards for others. In a direct line of inheritance from van Leeuwenhoek, global teams of researchers drawn from many different institutions successfully mapped the whole of the human genome, in three rather than the estimated 15 years. The

21st century mind not only has giant shoulders on which to stand, it has had many shoulders on which to lean.

Advances in ideas have always been the outcome of a collective mind, yet the inheritance of Descartes 'I think, therefore I am' has been to allocate all the glory to one leading figure. Denis Brian, in his 1995 *The voice of genius*, records interviews with 20 then-living science-based Nobel Prize winners. Individual characters though they were, they all spoke of the essential contribution to their discoveries of open-ended discussion with a few compatible spirits. These were not necessarily (nor even often) fellow geniuses. The prizewinners also recalled making their major discoveries initially through a creative leap in their minds before collecting the necessary evidence. The combination of strong social support and innate individual capacity was the key to their eventual recognition as genius.

Albert Einstein provides the model for genius for our era, as Leonardo da Vinci did for his. From his extensive journals, Leonardo appears to have been a social isolate, even a misanthrope. Yet his will reveals that he had lifelong friends and valued a life of conviviality. Einstein has been mythologised as a 20th century 'modern' scientist, an abstract thinker and a loner. On the contrary, he was also a violinist, a social activist and a practical joker. He shared with Brian's other geniuses the capacity to make the creative leap of the mind ahead of the evidence, and the need for open discussion. Biographer Ronald Clark in *Einstein: the life and times* reports that Einstein himself attributed his ability to produce his seminal papers at the age of 25 to the stimulus of a discussion group of fellow students.

Einstein was adapted for synthesis, not only intellectually but biologically. Having left his brain to science, investigators at first could only find that it had the same structure and weight as other brains. Eventually, Sandra Witelson, an anatomist from McMaster University, found that the *corpus callosum*, usually a distinct rope of nerve fibres connecting the right and left hemispheres (the source of our powers of analysis and synthesis), was a fused mass linking both hemispheres (figure 5.2). Einstein's genius in combining a vast grasp of mathematical structures with the capacity to make creative leaps arose at least in part from the close connection between the two hemispheres. He may well have had a greater physical as well as intellectual capacity for thinking synergistically, but as with Leonardo, the basic physiology and conditions of creativity are available to us all.

The idea that all humans have an innate capacity to arrive collectively at a new, integrative construction of knowledge has emerged in both sociology and psychology. Maturana and Varela give us the idea of 'autopoesis', the self-governing mechanisms by which we each orientate ourselves in the world. In education, Nowotny, Gibbons and others write of the new construction of knowledge based on integration and synthesis. In the biophysical sciences, a move to multidisciplinary

Figure 5.2 Connections in the human brain: the role of the corpus callosum

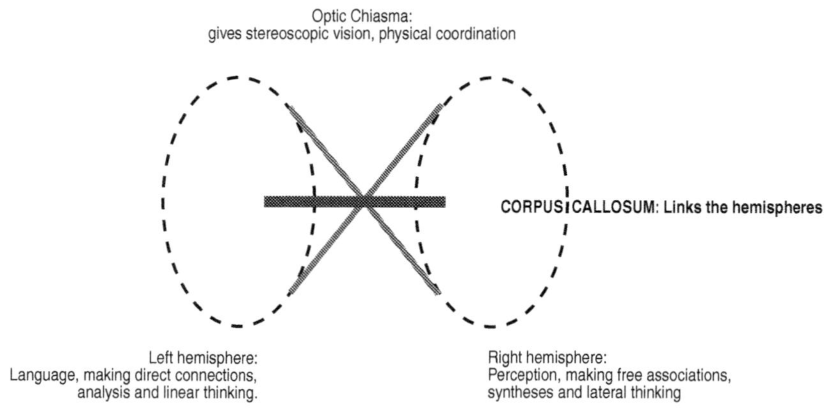

and systems approaches to understanding a changing planet has been labelled sustainability science, a move discussed further in chapter 7.

It would seem that thinking collectively is as innate as thinking analytically, and the swing of the pendulum is turning from one to the other. But in welcoming the change, we need to remember that this capacity is recovering from 10 generations of emphasis on analysis. As Smithson in *The changing nature of ignorance* predicts, protection of the dominant mode of thought has led to a general social denial of the capacity of ordinary human beings to think synergistically, to generate new thinking between them. People who try to move to collective thinking can be dismissed as too ambitious, and criticised for diluting their specialised skills.

Access to collective thinking tends to be widely underestimated. There are many synoptic methods for honing our capacity for collective thought, using analysis and synthesis, subjectivity and objectivity, individually and socially, locally and globally, and as individuals, community members, specialists, organisational contributors and holistic thinkers.

Donella Meadows in her 1972 *Limits to Growth* gave the world one of its first warnings of the global environmental disruptions from human activities. In *Vital Signs* in 1998, she quotes a Sufi story: *"You think because you understand one you must understand two, because one and one make two. But to understand two you must first understand 'and'."* (Donella Meadows 1998 *Beyond the Limits* p.ix). We go on to explore the significance of that 'and', as applied to the relationships between the personal and the collective capacity for synthesis.

From the example of acknowledged genius we have some principles to follow for developing collective genius, *what should be*. Genius is marked by the need for dialogue with sympathetic souls, not necessarily fellow experts in the same field. It is marked by the courage to make the creative leap, ahead of the evidence. And it is marked by the capacity to make use of the connections between analysis and synthesis, rather than either alone.

What is:
fragmented minds

Figure 5.3 Leonardo da Vinci *Head of a man with scheme of proportions*. Venice, Academia.

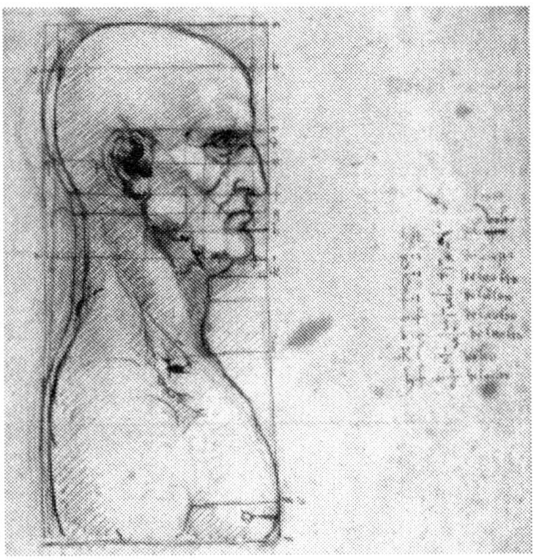

© Photo SCALA

Given Western culture's emphasis on dividing and classifying, we would expect to find that studies of human thinking divide the capacities of our minds into compartments. That is exactly what we do find. We have been conditioned to expect that we will have a bias toward lateral or linear thinking, simple or complex problem solving, emotional or logical intelligence, negotiation or confrontation, and individual, community, specialist or strategic knowledge. Following Donella Meadows' advice, we will try to turn those 'ors' into 'ands'. The following are the types of thinking into which our minds have been divided. All of these we can access if we wish.

Psychological gestalt

We start with the subjects ready at hand: ourselves. Figure 5.4 presents the classic test, not of intelligence but of the personal mode of synthesising evidence. Did you see the face in the figure immediately? After a few moments? Can you not see it at all? Studies confirm that about 5% of observers see the 'face' immediately, and after that cannot lose the image. About 5% will never see the figure, no matter how carefully it is explained to them. They will find this extremely frustrating. The other 90% will see it gradually, or after it has been pointed out. Everyone is intrigued by the task. There is no right answer. The figure is an ink blot.

Figure 5.4 Classic gestalt figure

Source: Abercrombie 1967

Whatever the individual form of response to the image in figure 5.4, in each case the biological sensation of pleasure drives the intellectual act of discovery. To test for yourself how much energy drives our capacity to make what psychologists call a 'Gestalt' (creating sense out of disparate evidence), try it out on your friends. Watch their response. And reflect on your own response to the figure known as 'Christ in the snow', as suggested in box 5.2. If you made a quick gestalt, you are accustomed to making sense of the evidence immediately. If you make a slow gestalt, you are used to becoming familiar with the detail. In between are a wide variety of individual strategies for sense-making, all an essential part of each person's identity.

Thinking strategies

The Greek poet Archilocus in the 7th century BC used the analogies of the fox and the hedgehog to describe two very different ways of construing the world. Quoting Archilocus, 'the fox knows of many strategies, but the hedgehog knows of one great and effective strategy'. Isaiah Berlin uses this in his seminal 1953 essay *The Hedgehog and the Fox* to distinguish between two classes of person: "*Those, on the one side, who relate everything to a single central vision, one system more or less coherent or articulate, in terms of which they understand, think and feel – and on the other side, those who pursue many ends often unrelated and contradictory, related by no moral or physical principle.*"

Berlin's essay classifies Tolstoy, Shakespeare, Aristotle, Montaigne, Erasmus, Moliere, Goethe and Joyce as foxes; they cover many themes, not necessarily connected in a single view of humanity. Plato, Pascal, Hegel, Dostoevsky, Ibsen and Proust he lists as hedgehogs; each is driven by a central idea. In our everyday lives, we can recognise both types of thinking, two sides of a collective understanding.

Knowledge-making

In a study of the learning strategies of 750 undergraduates in the faculties of Arts and Science and a cross-faculty program, the Human Sciences Program, students were asked to solve test problems, complete psychology personality profiles and write short scenarios. The results, published by this author in 1984 as *In formation: the place of specialists, generalists and holists in an information system*, revealed not two distinct learning strategies but three. The metaphor of thinking like a hound was added to Berlin's hedgehog and fox.

Many of the students could be recognised as fitting Berlin's metaphors. One group, enrolled in the interfaculty program with its goal of creating holists, fitted the criteria for hedgehogs. They were mainly concerned with understanding the core of issues. In solving problems and in dealing with social issues, they were looking to apply their learning for a useful purpose. They were intolerant of complexity, but tolerant of ambiguity and uncertainty. Another group, the specialists (hounds) most of whom were enrolled in science, were interested in covering each topic in detail and in following each trail of inquiry to the end. They were not primarily concerned if their knowledge was useful, but with whether it was correct and reliable. While they were tolerant of complexity, they avoided uncertainty and ambiguity. Students enrolled in Arts, and labeled the generalists, fitted the metaphor of foxes. These shared with specialists the intolerance of ambiguity, and with the holists an interest in their learning being relevant. Generalists sought a wide range of dissident views. They were more likely than the others to be intolerant of ambiguity and uncertainty.

These results suggest that personality types develop out of individual earning styles, and are encouraged by the organisational settings they chose. This reinforces David Kolb's work with organisations, where learning styles and occupations were strongly matched. An overall conclusion is that collective thinkers are both born and made.

Personality types

The 16-type matrix known as the Myer-Briggs Type Indicator offers an analysis of personality types that is frequently used in organisational development. The indicator cross-references the various ways in which people make their decisions, according to four dimensions: introversion/extroversion, sensing/intuiting, thinking/feeling and judging/perceiving. The personality matrix that results is intended to allow management to select a well-balanced team, well suited to collaboration. Each individual has the choice of extending his or her repertoire across the full spectrum, or making up part of a mixed team.

Styles of conflict resolution

In a study of management styles, Hugh Watson identified four types of response to conflict, each appropriate to some situations and counterproductive in others. They are avoiding conflict, forcing a solution, smoothing and compromising. The four styles coming together without a mediator can escalate a dispute to stellar heights.

Box 5.2

TESTING: A PERSONAL APPROACH TO KNOWLEDGE MAKING

The information in figure 5.5 sets out a series of personal profiles of knowledge making. Consider each of the dimensions in the table below. Were you fast, medium or slow at arriving at the Gestalt? Do you currently use all the available avenues for creative thinking, problem-solving, experiential learning and knowledge cultures when presenting an argument, or do you favour only one?

Is there a consistent pattern in the way you approach the knowledge-making task, or do you favour ways of analysis for some thinking tasks, and synthesis for others? Are you tolerant or intolerant of uncertainty and ambiguity?

Your quick self-diagnosis can be followed up in more detail through specialised diagnostic tools available through the texts of the relevant authors, or accessing through Google. The purpose of the exercise is for you to recognise what forms of thinking you DON'T use. Using the list in this section and the chart in figure 5.5 ,you can remedy this by extending your repertoire.

The fifth response, a collaborative problem-solving approach, is the key to conflict resolution on complex issues. Each participant diagnoses their own conflict resolution style in concert with the others. This gives a transparency to the process which opens up the capacity for dialogue, no matter how heated the debate. The rules of dialogue then allow the participants to hear each other, opening up an opportunity to invent new solutions to old conflicts.

Multiple intelligences

Howard Gardner, in his 1983 *Frames of Mind,* challenged the orthodox assumption that all intelligence is rational and objective. He identified eight types of intelligence and suggested there well may be more. The intelligences - logical, mathematical, linguistic, musical, interpersonal, intrapersonal, spatial and kinesthetic - have now all been the subject of field studies, reported in 1993 by Gardner in *Multiple Intelligences*. Confirmed by many educational writers since, the types of intelligence have been allocated to certain occupations and personality types. In practice, we use all of them all the time.

Organisational context

We have already discussed in chapter 4 how David Kolb found that particular occupations developed one stage of the learning cycle more than others. Researchers, scientists and engineers had a talent for convergent thinking, moving from the general to the particular. Artists, writers and poets displayed divergent thinking. Service professionals, such as in health and administration, were strong at assimilating, linking new ideas to an existing framework. Organisational managers and theatre directors were likely to

Figure 5.5 Twin forms of knowledge making: analysis and synthesis

Source:	Dimension	Preference for analysis	Preference for synthesis
Gestalt	Perception	Slow gestalt	Fast gestalt
Berlin	Thinking strategies	Fox: seeking to know many things	Hedgehog: seeking to know one most important thing
Brown	Knowledge building	Tolerance of complexity Intolerance of ambiguity Hounds and foxes	Tolerance of ambiguity Intolerance of complexity Holists
Kolb	Learning styles	Convergent thinking Assimilation	Divergent thinking Accommodation
Gardner	Multiple intelligences	Logical, mathematical, linguistic, musical	Interpersonal, intrapersonal, spatial, kinesthetic

be accommodating, changing the framework to adapt to the new ideas. In commonsense terms, however, everyone uses all of these skills some of the time. Some of the most spectacular advances in introducing social change have been the work of people who combined all four.

What could be:

embracing all the knowledge cultures

Consciously or unconsciously, we are already using all the knowledge cultures and all the intelligences in making decisions, large and small. Under normal circumstances, much of that thinking activity will remain unexamined, within the tacit knowledge so well described by Polanyi. Decision-making based on synthesis requires the further step of doing this consciously, reflecting on our learning. We can do so through several routes, as disciplinary specialists, as members of a synthesis team, or as expert synthesists in our own right. Box 5.3 presents an example of how one individual constructs an argument by drawing on all five of the knowledge cultures.

In the letter to the editor on genetically modified organisms, the Australian environmental educator Frank Fisher draws on his personal experiences in controlling his environmentally-based Crohn's Disease, his understanding of his society, his skill as an environmental scientist and educator, and strategic ways of influencing government. His holistic focus remains on a sustainable future for both natural systems and the human species. His conclusion is that for an individual to be scared of socially-driven biophysical risks to global integrity, such as genetically modified organisms, even if scientifically unproven, is a reasonable application of individual common sense.

What then is the significance of identifying these multiple ways that humans can order their thinking and construct their knowledge? First, it establishes the versatility of the human mind in responding to the challenges of sense-making, mounting a convincing argument, creative problem solving and dialogue. It also highlights the importance of recognising the relationships between parts and wholes in human intelligence, and of closing the gaps in collective thinking within the repertoire of each individual.

With this heightened understanding, we can turn to the theoretical perspectives and practical experience that will help link each individual's own synthesis, and their contribution to the collective mind.

For the 400 years of the scientific Enlightenment, Western society has recognised advances in knowledge by rewarding a few specialist individuals. It is becoming more usual to award Nobel Prizes or research scholarships to multidisciplinary teams. It is still unheard of for an award to recognise membership from all the knowledge cultures, however central these may have been to the work concerned.

For instance, in 1999 a social worker identified a member of her AIDS support group who after 15 years had not developed full-blown AIDS from its precursor infection HIV. She traced their contacts, and discovered the same story wherever the infection had been

Box 5.3

A LETTER TO THE EDITOR: BEING SCARED OF GM

Mr. A. Howe, Editor, *Sunday Herald Sun*,
Melbourne, 30 December 2002
Re: Scared of GM? An environmental scientist
responds to "GM scare hots up",
Sunday Herald Sun, 29.12.2002, p30

Dear Mr. Howe,

"Being scared of GM" may not be rational but it is reasonable. Here's why. Humans are still very much taken by their apparent power over nature. The insights of science have heightened both the scope and extent of that power and genetic manipulation represents a quantum leap in both.

Now, science is not itself power. It is the careful attempt to theorise and build insight that stands the tests of repeated experimentation and open criticism over time and varied practice. The creation of theory is the domain of inspiration. Transforming theories into science, however, is the domain of rationality. It involves finding and running experiments that fit into what is already accepted as science and then subjecting the results to repeated criticism. In this effort, science is our most noble creation and so the SHS is to be congratulated for running its news article on poor science teaching in schools (29/12, p18).

For all that nobility, science is not and never can be ultimate truth. It is the most reliable set of interpretations we have at any time and is, by the nature of science, always open to questions about its insights (laws and accumulated details) and methods. We definitely are not in a situation to say we have "nothing to fear from GM foods". We can listen when a proponent of GM foods says "GM ingredients have been on our supermarket shelves for six years without a single scientifically reputable report of any adverse impact on human health". As a scientist myself I must have misgivings about how well such a proponent knows the field of writings on the topic and, if he does know them, how he is interpreting those writings.

A more important concern with GM foods is one that goes way beyond the direct health implications to humans. It arises from the doubt we must always hold about scientific knowledge and the contexts within which it is applied. These are reasonable concerns but not rational ones; one cannot substantiate them with science because both come from outside science and its apparatus of proof. They arise from our personal and social interpretations of the case.

We are sufficiently powerful to suppress nature's attempts to reject us. As a still-living sufferer of an auto-immune disease (Crohn's) I am a living example of that power. We have become ourselves a global or nature-wide influence. This in itself may not be a problem; but we have now established world-spanning systems that make it difficult for nature as a whole to protect us.

With only the shallow government assessment structures we currently have to judge what the market presents to us as GM foods, we are determining the future of something whose implications are very broad. And again, while this is not new - indigenous Australians transformed the continent with fire - we are currently wilfully disregarding the importance of these systems to our everyday lives and to the priorities that dictate what they should be. Of course, nature will prevail, but it may do so in ways that are unpredictable to us and we may not like them. Indeed much of nature may not like them either!

Yours sincerely,

A/Prof. Frank Fisher
Director, Graduate School of Environmental Science
Monash University

Source: Frank Fisher, Monash University, quoted in Brown et al 2005 p137

transmitted from the same contact. This attenuated strain of AIDS promised to provide a source of immunisation against the infection. The social worker advised her research team, and medical researchers went on to publish to considerable acclaim. There was no recognition in the publication of the social worker, nor of the administrative staff that kept the meticulous records that allowed the original discovery. And there was no recognition of the HIV sufferer who was the source of the original information.

If we are to move to modes of inquiry that are inclusive rather than exclusive, a celebration of the diversity of the contributors will be crucial. Rather than maintaining the polite fiction that only the nominal team leader deserves recognition, there needs to be exploration of the web through which new knowledge is generated, as in the AIDS example above.

A range of sociologists, such as Peter Wagner in *Sociology and Contingency*, Bruno Latour in *We Have Never Been Modern* and Michel Foucault in his *Archaeology of Knowledge* agree on the three new capacities that set our era's constructions of reality apart from what has gone before. As Bohm recommends, all three depend on celebrating difference, rather than looking for a single ideal. All three need to be incorporated in a collective view of the world.

First, we are the first generation to accept that our knowledge has no reference to a final authority, but is contingent on the social *Zeitgeist* and the interpretation of the biophysical world.

Second, our vastly increased technical capacity has led to the realisation that we as human beings in all our frailty, not some external physical or metaphysical force, hold in our hands the options for how the world will be. Third, we are the first generation to knowingly harness the human capacity for reflexivity; that is, to reflect on the changes brought about by our own learning, and those of our fellows.

I-It, I-Thou, I-You

In trying to understand the effect of these changes, we have so far reviewed psychologists' divided perspectives on the individual mind, and sociologists' on the current state of the collective mind. It is a philosopher who provides us with the connecting link between the two. Martin Buber, a German-Jewish philosopher of the 1930s, has reflected deeply on the relationships between the individual and the collective. His advocacy for the power of dialogue predates Bohm's practical rules of dialogue by half a century, and follows those of Socratic dialogue by a millennium.

Buber takes the position that the mind of each individual can only develop through his or her interactions with other humans. We know this to be technically true, since psychology studies from Piaget to John Bowlby confirm that no human can learn to speak, love, hear or reason without contact with other humans. Buber, however, refers to something equally fundamental. He concludes that humans structure their identity and their relationships with reality on the basis of how they structure their human-to-human relationships.

These he differentiates into I-It, I-You and I-Thou. The I-It relationship is an inheritance of the industrial era, when people began to use machines to undertake previously human tasks. Charlie Chaplin powerfully demonstrates this mechanical construction of human beings in the film *Modern Times*. Partners are not equal in the I-It relationship because it implies ownership of one by the other. However, the I-It dialogue cannot be simply discarded, since it forms the basis for objective knowledge and of the power of specialised forms of knowledge. In I-you relationships, people acknowledge each other through the social rules that bind them, with formal greetings and language.

It is possible for humans to place themselves completely inside a relationship with other humans, to truly understand and 'be there' with another person, without social masks, even without words. Such a relationship Buber calls the 'I-Thou'. In psychology it is known as empathy. In Buber's use of the term, each person comes to such a relationship without preconditions. The bond thus created enlarges each person, and each person responds by trying to enhance the other person. The result is true dialogue, a sharing of minds.

In the field studies that led to this book, there are examples of these qualities in practice. We now examine some examples that emerged in the course of developing the skills of individuals' integrative inquiry.

What can be:
the spirit of integrative inquiry

In 1996, I had the opportunity to establish a post-graduate program in an applied science faculty. This meant developing a synergistic learning process within the heart of the linear-thinking, analytic tradition. The theme of the program was *Managing for local sustainability*. The course was dialogue-based, and set in the decisions-into-practice framework. The program provided a rich set of contrasts between the role of the individual in the specialised and collective process of inquiry. The criterion for a successful Doctorate of Philosophy (PhD) is a fresh contribution to knowledge. Traditionally, a PhD thesis has been an intensely individual process. The learning process has been authority-driven, in the master-disciple mode. Candidates have been expected to make objective observations, limiting their conclusions to 'what is'. In this program, the students formed a learning circle, in which each contributed to the other's learning, and course interactions were based on developing collective thinking.

The formal acknowledgement of success in a PhD program is the number of post-graduate candidates who gain the degree. In this case it was 10 out of 17. Each of the 10 had been a participant and, in five cases, an instigator of the very change processes they were examining. Equally important for understanding the role of synergy in the inquiry process were the seven students who did not complete their degree. Two found they preferred to stay with analytic thinking and moved back to specialised departments. One withdrew to write a book and one to produce a video, as better suited to their individual creative goals. Two failed to complete the writing-up of their research results, life events having made the complex task too difficult. One suspended enrolment, saying they needed more life experience before they could manage the broad perspective of the program.

The program broke several cardinal university rules. The research students were drawn from across departments, rather than all coming from one. Their topics were generated from personal experience, rather than the needs of a discipline. There was one primary supervisor, myself, for all 17 students, when there is a usual limit of five. That single supervisor was not, as is normally expected, an acknowledged specialist on all of their topics - or even on most of their fields - but was experienced in the theory and practice of adult learning and the management of social change.

Mutual learning processes helped the group turn inward, supporting each other with a strong sense of critical loyalty. Their commitment to their inquiries gave the group a driving energy. An external evaluator, Jan Malfroy, in her own PhD thesis noted the synergy as rarely duplicated in traditional programs. Each graduate included the seminar and its members in the acknowledgments of their thesis. This commitment and mutual support meant that a single supervisor could support the individual student and the group collective in their learning processes, leaving the required specialised expertise to specialist advisors.

Synergistic seminars

Two synthesis-based processes evolved during the development of the PhD program, a synergistic seminar and a synoptic mode of inquiry. The first was aimed at developing creative thinking collectively among the graduate students, generating energy from the interaction. The second was a research design that enabled an investigation into a wicked problem, by accessing the multiple dimensions and incorporating paradox.

The synergistic seminar based on mutual learning was held one evening a month. The guiding principles were critical loyalty and the rules of dialogue. All participants contributed equally to the critical evaluation and extension of each other's work. The non-negotiable rule was for participants' contributions to assist each other, and not attempt to prove a point or establish superiority. All participants (including the supervisor-convenor) had equal right and equal responsibility to set the agenda, contribute and comment constructively on each other's contributions.

Conversation followed the Bohmian rules of dialogue as outlined in chapter 3, as opposed to the expert-based hierarchical rules of the traditional seminar. Key terms, such as transformation, sustainability, community and integration, were not narrowly defined, but expanded, using the personal construct mapping tool. A basic tool in the seminar and research process was Open Space Technology, developed by Harrison Owen in 1992. Four principles and one rule establish an open learning space, waiting to be filled by whoever chooses to be involved. The entry requirements are a passion for and commitment to resolving a wicked problem.

The Open Space Principles are as follows:

- Whoever comes are the right people;
- This time is the right time;
- This place is the right place;
- Whatever happens is the only thing that could happen; and
- When it's over, it's over (in other words, do what you feel is of value, and when it's done, move on).

The synergistic conditions for the seminar included the PhD students being the right people, strongly committed to their inquiries. The right time was 6-9pm once a month, over a shared meal. The group was often still talking at midnight. Sub-groups met in between, on their own initiative. Dialogue themes for the next month were chosen at each monthly session, generated from that week's discussion and the needs of individual participants. The right place was a round table in a pleasant room.

In spite of the explicit agreement on the use of dialogue, this did not avoid the need for facilitation, and even adjudication. There were issues of individuals' dominance of the discussion. Some members originally took little part, and others found it difficult to avoid the use of specialised language. A serious issue was the need to trust. Some of the group had been wounded by pre-emptive specialist criticism. A difficulty sometimes arose with the contributions of visiting academic staff who were still working within a specialist critique.

Synoptic inquiry

The second mode of synthesis developed within the PhD program was a mode of synoptic inquiry that included all five knowledge cultures. Standard research follows the sequence of a key proposition (the thesis) - defining its opposite (the anti-thesis) - the evidence or argument for and against - and the resolution of the argument/ conclusions (synthesis) (figure 5.6).

In synoptic inquiry, the research design follows the decisions-into-practice learning spiral described in chapter 4. Starting with *what should be*? as the proposition for the thesis, the second step, determining *what is*? is the diagnostic stage. Here the key variables are decided for the wicked problem that treatens the proposition on *what should be* (antithesis). Only then can the inquiry turn to what could be: the path of the most fruitful social change towards what should be. Testing *what could be* in the actual world turns the thesis into a science thesis as wel as a collective one. The choice of inquiry methods (observations, interviews, intervention or experiment) depends on that decision. This is a major point of departure from classic specialist inquiries that design their inquiry around the tools traditional in only one discipline.

The core of sustainability decisions is reaching a collective understanding, drawing on the full set of nested knowledges. This multiple evidence

Figure 5.6 Steps in a synoptic inquiry e.g. sustainable cities

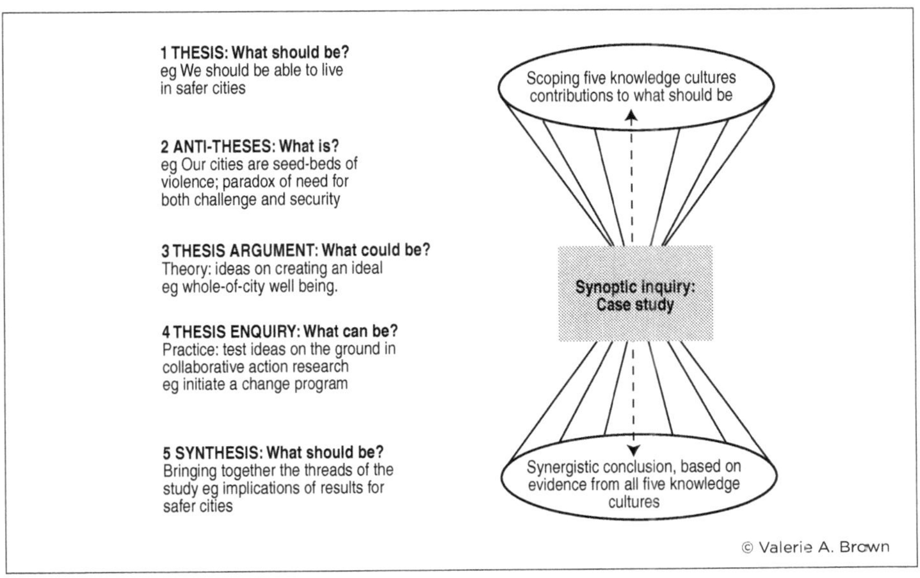

base applies to both the thinking of the individual researcher, and to the subject of their research. The outcome has been found in practice (and in the eyes of thesis examiners) to be more insightful, and closer to practical application, than any one of the knowledge cultures or their representatives could contribute alone. The following theses were produced in this way.

Brian Dwyer, finance manager and engineer, titled his thesis *Aspects of governance and public participation in the remediation of the Murray-Darling Basin.* His study site was a small country town, and his research method was first-person observation, as from a 'stranger' in the town. He recorded his observations as vignettes of the discussions he had with the people there, and events in which he took part. Dwyer's findings were that social change had four legs: the local style of governance, the form of the local landscape, the personal construction of reality of each citizen, and the roles played out in the community. Among other insights from this approach was the discovery that in any community there are individuals who act as agents for change who influenced all four of these dimensions of governance. Dwyer called these people '*nexors*', derived from nexus, a meeting point, a point where ideas and people combine.

Rod Griffith asked *How shall we live? The sustainability agenda and institutional change in local governments in Australia.* A senior environmental planner with 15 years' experience, Griffith adapted the decision-into-practice framework to fit local government practice. He then selected nine councils that were committed to sustainability. Griffith found that in practice there was a very mixed understanding of what sustainability meant. Local sustainability proved to need a major and concerted re-orientation of council processes, from style of senior management to personal relationships between staff of different departments. From the change agents who stayed and from those who had gone on, the consistent reports of stress and burnout indicated this was an important area to redress.

Box 5.4

SOMETHING TO THINK ABOUT: FAITH IN THE INDIVIDUAL

"Our deepest fear is not that we are inadequate. Our deepest fear is that we are powerful beyond measure. It is our light, not our darkness, that most frightens us. We ask ourselves, who am I to be brilliant, gorgeous, talented, and fabulous? Actually, who are you not to be?............ Your playing small doesn't serve the world. There's nothing enlightened about shrinking so that other people won't feel insecure around you. We are all meant to shine, as children do............... It's not just some of us, it's in everyone. And as we let our own light shine, we unconsciously give other people permission to do the same. As we are liberated from our own fear, our presence automatically liberates others."

Marianne Williamson 1992 *Reflections on the Principles of A Course in Miracles* Harper Collins

Mazrura Sahani is a medical practitioner, researcher and Malaysia's expert on the haze (air pollution from forest burning across the region). The goal of her thesis was to find ways for communities to increase their capacity to respond to environmental pressures such as haze. Her inquiry involved five Malaysian communities: an urban community with government support; an urban community not so supported; an ethnic minority community; a group of medical specialist advisors; and a group of specialist environmental health researchers. Mazrura found that the responses from the residents inside, and the helpers outside these communities were so different that they appeared to inhabit separate realities. Solutions required the collective collaboration and open communication among them all.

The lessons from these and other theses of the Local Sustainability Project were that social transformation begins with committed individuals. Everyone has the capacity to draw on the five knowledge cultures, use all the problem-solving and thinking styles and work in close personal trust relationships with others. But there is still a need for a setting which allows synergy to be achieved. Valuing difference, contributing as an individual to the collective mind, and reflection on the full circle of learning, are strengths in seeking to guide social change. These are the intellectual dimensions of thinking individuals, just as much as Leonardo's drawing of the so-called Vitruvian man gives the physical dimensions (figure 5.7).

Figure 5.7 Leonardo da Vinci. *Scheme of the proportions of the human body* (or Vitruvian man). Venice, Academia

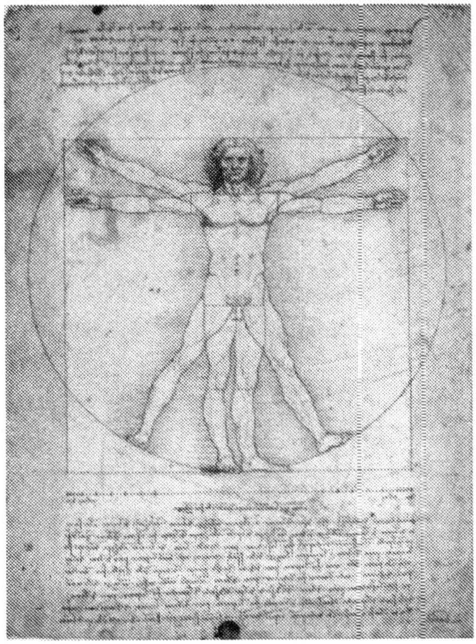

© Photo SCALA

In proceeding towards a sustainable future along the route of nested knowledge cultures, whether the program is large or small, the first step is the commitment of the individual to achieving collective thinking. The next chapter takes the next step in the nested knowledge cultures, and adds local community knowledge to guide collective social change. .

6 Community engagement

Q. What is related to time as place
is related to space?

A. A story

A story is to time
as a place is to space
And a place is a story told many
times.

A koan, collated from Ralph Metzner *Green Psychology*

Dreams of an ideal community have always been grounded in stories set in a particular place and time. This holds for the great missionary religions of Buddha, Christ and Mohammed, who promised Nirvana, Heaven and Paradise. Long journeys took the Pilgrim Fathers to the United States and the French Acadians to Canada in search of their visions of the future. Less happy attempts include the tragic tales of Waco and Jonestown. Here the leaders eventually sacrificed the disciples. More positive examples are Henry Thoreau's 1846 back to nature at Walden Pond in New England, and the futuristic use of self-sufficient technology at New Eden in Britain. In each case the unit of social change has been whole-of-community engagement in a shared vision. The direction of the change has been set by new stories carrying symbols of the shared vision.

What should be:

shared symbols of people and place

The potential for whole-of-community learning rests on a paradox. Any lasting change in any one community depends on a matching transformation in the surrounding society. Any lasting change in the larger society depends upon transformational change in its constituent communities. This paradox leads to the wicked problem of where does change begin. Like much of the shift in adding synthesis to analysis, the answer lies in welcoming the paradox, not reducing it to either/or. Communities and their larger society are so strongly interconnected that they will catalyse change in each other. All members of a community and its encompassing society are parts of a single learning system.

Communities come in all shapes and sizes. They may be long-standing – embracing a physical site, knowledge, values or skills – like the great cities of Paris and New York. They can be created spontaneously through shared life events, as in an immigrant community or one that recovers from disaster such as a flood or bushfire. And they may be intentional, a chosen way of life, as in monastic, religious, communal and self-sufficient communities.

Community is as much part of a human's sense of reality as the pod is to whales, the pack is to wolves and the flock is to birds. Shared rules and meanings hold individuals together. They can be disrupted but never entirely banished from their minds. The regular pattern of return to their original family villages by sophisticated city dwellers in Europe and China whose ancestors may have migrated three generations ago reflects the strength of that community drawing power.

The many predictions of the death of community have proved premature. Canadian sociologist Jane Jacobs came to the defence of the protective self-organisation of urbanised communities in her 1961 book *The death and life of great American cities*. She broadened her scope in 2000, with *The nature of economies*, to argue for each community to become aware of its global connections. Robert Putnam is a seminal writer on social capital and the human relationships that hold a community together. In 1995, he wrote an article on *Bowling alone: America's declining social capital*. By 2000, he had moved on to publishing a book entitled *Bowling alone: the collapse and revival of American community*.

At any one time, individuals form the central link in the chain of five generations: our parents, their parents, our own generation, our children and their children. 'Living dead' traditions are handed down, often without direct knowledge of the source. Freud's heir, the psychoanalyst Jung, describes the way in which communities develop their own collective reality through shared events and experiences.

City-dwellers perpetuate that reality through powerful symbols that hold the community together between generations and over time. Renaissance thinkers designed perfect aesthetic spaces, waiting for the humans that would bring them to life. The balance of space and building in Leonardo's design for a city in figure 6.1 is an example. New York's Statue of Liberty commemorates the great 19th century influx of immigrants. London's Houses of Parliament stand for 600 years of parliamentary government. Berlin maintains the burnt church of the Kurfstendam as a symbol of defeat and resurgence from World War II. Sydney has its Opera House, designed as sails floating on one of the world's great harbours. Smaller cities and settlements have equally powerful symbols of their own, instantly recognised by all

Figure 6.1 Leonardo da Vinci *Plan for the city of Imola*. Vinci, Museo Vinciano

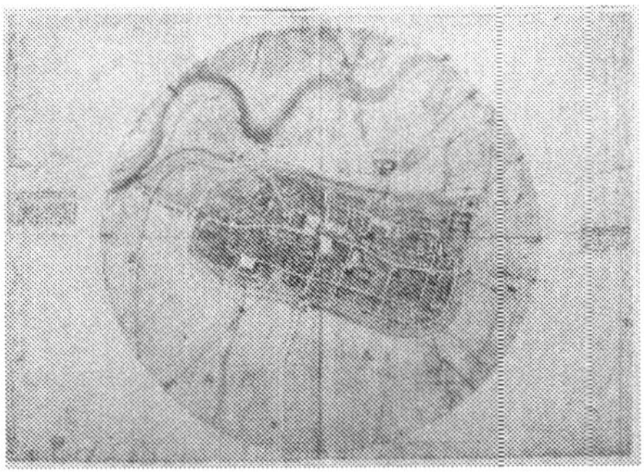

their residents, and often quite mysterious to outsiders.

Inquiry about the symbols of community identity can give insiders and outsiders the key to the essential drivers of that community. Lewis Mumford, a writer famed for his capacity for synthesis, points out in the 1961 book *The City in History* that three essential public symbols represent every human settlement from the earliest nomadic camp to a modern megalopolis. In the earliest settlements, archaeologists have found sites for safeguarding the leadership, preserving the sacred, and storing resources against want. Mumford identifies these as a palace, a temple and a granary. Together, their presence generates the mutual sense of identity, spirituality and security crucial to the human sense of place.

Recent experience confirms that Mumford's propositions still hold. Over the past few years I have led many workshops for people seeking to strengthen their sense of community. A synoptic workshop process has developed, based on the three symbols of palace, temple and granary. The results of one such workshop are presented in figure 6.2. A longer account has been published as *The symbols of my place are the foundations of my world*.

The diverse participants came from all the knowledge cultures, and identified the three symbols in their many different ways. Yet the symbols were always linked into a whole. For example, one country dweller identified the farm fences as forming the palace which was the land itself, the homestead as the temple and the herd of stock as the granary. Another country person nominated their children as the granary, reflecting a strong sense of the future. An academic nominated her library as granary, her university department as palace and the seminal authors in her field as temple. Different individuals offered the national Parliament House (the largest building built in the southern hemisphere this century, and a focus of national design and crafts) as palace, temple and granary. Anyone from inside or outside this community could get the 'flavour' of the community from these responses.

This exercise can be repeated for any community, any place. Lessons drawn from a range of such exercises include:

- everyone has a deep and abiding individual sense of place as part of their own identity;
- everyone can recognise the ways in which a place meets their physical, social and spiritual needs, expressed in a great variety of symbols which form patterns of shared understanding;
- the same structural component of place can be valued by a whole community, while representing very different values and functions to different groups and individuals; and

Figure 6.2 Site-seeing in the capital: interpretations of symbols of Canberra, Australia

Symbol:	**Interpretation of symbols:**
Palace: Site of social safeguards and governance	**As palace:** Parliament House, town plans, home, National Art Gallery, High Court, Public Service Board.
Temple: Site of spiritual power, ideals, wonder	**As temple:** Parliament House, the bush, hills and lake, Botanic Gardens, War Memorial, hospital, churches, swimming pools, roads, night sky.
Granary: Site of resources stored for present and future needs	**As granary:** Big W, McDonald's, shopping mall, libraries, Treasury, sales, vegetable gardens, banks.

- a sense of place and community is built from the participating individuals out, and can never be understood in the same way from the outside in.

So shared symbols provide the common language that binds a community. Writers such as Robert Putnam interpret the power of these symbols as providing social capital. The idea has been expanded so that Friedman and Putnam write on bonding and bridging social capital. The first refers to the social connections that bind groups together, and the second to those connections that link a community to other groups. Human ecologist Rory Eames, in his article *Partnerships in civil society: linking bridging and bonding social capital*, writes of the strategic social capital that crosses between the two. In the examples of whole-of-community change that follow, we discover just how strong the bonds of inside-out and outside-in can be.

What is: **fragmented strategies for social change**

Around the world, the relationships between communities and their surrounding society are caught in dramatic discontinuities from the past. First, withdrawal from global influences is no longer possible. The United Nations' state-of-the-environment reports confirm that there is no part of the planet's surface, from Siberian forests to the Antarctic continent, unaffected by human industrial activity. In the past 50 years the entire human population has become as strongly interconnected as the global natural systems, and large inequities are putting great strain on physical and social systems.

In the second discontinuity, communities are no longer able to provide a sustainable enclave free from the reach of large cities. The proportion of the world's population living in cities is more than 60% and rising. The ecological footprint of a city – that is, the amount of productive land needed to support the population – is seven times the actual physical footprint, according to the calculations of the concept's originators, Bill Rees and Mathias Wackernagel. The city reaches out far into the landscape for its resources. And every rural community draws on city services in some way.

Third, the dream of the perfect community is fading. Ironic suggestions for the ideal urban community now include a giant shopping mall, a satellite in space, or a drug-mediated nirvana. Community planning is polarised between the goals of the free market or the fully planned future, high technology or natural systems and being either human-centred or god-directed. Once again, we need to change the 'or' to an 'and'.

Fourth, a community is no longer strongly bounded. Before the era of global information and resource flows, palace, temple and granary worked as one. Territorial boundaries guarding local resources (the granary) were locally defended. These boundaries are now permeable, letting community members out and foreign influences in. Richard Falk, writing on global governance in *Predatory globalisation*, has identified what he calls the four sociodramas of globalisation-from-above. These are the vast global flows of finance, information, population and resources that sweep around the planet.

Developing separately is globalisation-from-below. Networks of communities, with their bonding, bridging and strategic social capital, make up civil society in their own right. The typical community is no longer localised, rural, authoritarian and closed. It is globally connected, distinctively urban, pluralist in its internal diversity, and open to global flows of finance, information, people and resources - an island dependent on its surrounding sea.

The very idea of community knowledge is frequently trivialised. Government agencies and the professions tend to apply the label disparagingly to ordinary citizens' knowledge. Consulting with 'the community' often refers to speaking to those groups who are uninformed, marginalised or powerless. In applying this weak interpretation to community consultation, members of government agencies, industry and the professions speak of community members as population, consumers and clients. They forget that they, too, are part of the community being consulted.

Local knowledge is constructed among people sharing a relationship with each other over time. An outside observer can convey the demographic profile and report on subjective interviews. They cannot, however, convey the unique quality that comes from being a part of that community.

In discussing local community knowledge, Indigenous knowledge is a special case. 'Indigenous' customarily refers to those communities who have lived traditionally in partnership with their land over centuries. 'First peoples', as used for American Indian tribes who established the initial human/environment relationship in their regions, is sometimes a more appropriate term than indigenous. This usually excludes Western traditional settlements. 'Native to this place' is a phrase more likely to be used to describe those born in and long resident on their land, whether first peoples or later immigrants.

At its most practical, a community is a set of people who are brought together by choice or force of circumstance, and who have learned to live, work and play together. Adapted from *Community development* by Ife, box 6.1 holds ten markers of a vibrant, continuing community. Each marker involves a synthesis of one dimension of community.

Kelman writes of three levels of being involved in community. The first occurs through compliance, formally accepting community rules and behaviours, as in signing on to a political party or a bushwalking group. This requires some degree of commitment and mutual trust. Later comes identification, adopting the ways of the community as one's own behaviour and values. This requires a greater degree of confidence, commitment and trust. Finally, a community member may incorporate their community's constructions of the world into their own identity and sense of reality, in a process of synthesis and internalisation. Members no longer question their own local loyalties.

Box 6.2 contains an account of a serious effort to establish a sense of community between urban and rural dwellers, farmers and local industries, government and local community residents

Box 6.1
BUILDING COMMUNITIES THAT LAST

Ife lists 22 markers by which a community is built and maintained from within, summarised as:

- Collective development: Community-building among the knowledge cultures of individuals, local community, specialists and organisations involved.
- Equity: Differential advantage is endemic to a community, and so there are local processes for adjustment and compensation.
- Sustainability: Environmental and economic resources are finite, so optimising both sets of resources is basic to community sustainability.
- Power relationships: The tensions between the personal and the political and individual and public issues are essential in any community.
- Belonging: A community's sense of identity lies in its collective sense of belonging between people and place.
- Goals and visions: Vital to a community's existence as defining a shared future.
- Organic development: A community is essentially organic (plant-like), rather than mechanistic (machine-like).
- Uniqueness: Particular set of attributes of people, place and time.
- Inclusiveness: All members are intrinsically part of the community even where they hold dissimilar views – every outcast marks a boundary.
- Pattern: A sustainable community contains a diversity of interests, ages, constructions of reality, learning styles, and personality styles, weaving a complex pattern.

Source: Adapted from Ife, 1995 *Community Development*

involved in managing the same vulnerable landscape. The example demonstrates how successful a synoptic approach can be, and at the same time how vulnerable the process is to sabotage by claims to dominance from any one of the knowledge cultures. In this case, it is the government host that undermines the process, but the other sectors can sabotage too. Local communities are extremely cautious about sharing their knowledge and so their power with the other sectors. Specialists are proverbial in being unwilling to open up the language which protects their 'secret' knowledge and so their power.

Box 6.2

COLLABORATIVE CONFERENCE FAILS

A national conference on the future of agriculture in a primary-producing country sets the context for a synoptic workshop. The conference brought together farmers, graziers, rural commodity dealers, environmental scientists, ecologists, geologists, political scientists, social scientists, urban planners, federal, state and local government agencies and non-government organisations from social, economic, conservation and rural industries. By the close of the conference, talk had become wide-ranging and open. People were listening to each other. A great sense of momentum had emerged.

Participants agreed during discussions that, yes, we have agreed on a new direction. Yes, it is a matter for city and country alike since the cities are sponges for rural resources, including urban sprawl over productive land. Yes, we may hold different positions on the solutions, but we can agree to work together on finding a common way to go. At the final plenary, panel members from a range of interests endorsed the agreements. And what happened next?

The rapidly expanding bright bubble of inventiveness burst. An enthusiastic farmer asked from the floor how to go about formalising the plans for future collaboration. The conference chair, from a government agency, answered: "Motions aren't appropriate here. We aren't an industry lobby." The rural participants fell quiet, not feeling sufficiently confident to push their cause. The specialists took over, each concerned for their own particular areas of interest: "We need more funds for weed research"; "Everyone should apply the findings of our water study". No more discussion, simply statements of different interests. People straggled out, muttering their disappointment.

The debacle was nobody's 'fault'. The system of boundary-maintenance had simply re-imposed itself, reinforcing the traditional knowledge hierarchy, and breaking apart the tentative formation of a knowledge network. At this crucial point, there was no affirmative action designed into the meeting to ward it off.

Source: Fenner Conference, Canberra October, 1999

What could be:

negotiating between islands and beaches

The Renaissance cities of Italy, where the integrity between people and city design is still functional five hundred years later, are proof that sustainable communities can be created and maintained.

However, in seeking to understand whole-of-community change in the present era of globalisation from above and below, we cannot simply return to a very different past. One useful perspective is to combine theoretical perspectives from anthropology with practical applications of the role of nested and networked knowledges in social change. The lessons of Greg Dening's *Islands and beaches: discourse on a silent land, Marquesas 1774–1880* help to explain the processes of a successfully coordinated whole-of-community change towards sustainability.

Dening writes: " *'Islands and beaches' is a metaphor for the different ways in which human beings construct their worlds and for the boundaries that they construct between them. The islands and beaches are cultural rather than physical. They are the islands men and women make by the reality they attribute to their categories, their roles, their institutions and the beaches they put around them with their definition of 'we' and 'they'.*"

We can make use of Dening's imagery to think of each community as a social island that is set within the knowledge structure of the wider, dominant society. Each island is surrounded by the unclaimed territories of the encircling beach, across which social learning can be negotiated. Each community has internal strengths developed from shared palace, temple and granary.

A three-year government-sponsored green house gas abatement program, Cool Communities, provided a chance to study 22 communities incorporating whole-of-community change. The program assumed that each community was a social learning unit, and a community's reconstruction of its knowledge was the essence of social learning. The following story of the change processes is summarised from *Social learning in environmental management* edited by Meg Keen, Valerie A. Brown and Rob Dyball. The influence of gas emissions from energy use on climate change is discussed in box 4.7 of chapter 4 of the present volume.

The 22 communities ranged from a football club in urban Melbourne, Australia, to a remote indigenous community, a church parish, a timber and agricultural town, a group of research scientists and a cohort of university students. Each community joined their industry, government and citizen groups to create their own Cool Communities project. The commitment required was to rise to the challenge of helping the world live sustainably.

Questions posed in studying the program included:

- Are there particular roles that need to be played by individuals in this particular whole-of-community learning process? (individual knowledge).
- How do the citizens find a valid communal voice in whole-of-community decision-making? (common sense or local community knowledge).

Figure 6.3 Knowledge cultures'* role in cool communities (n = 22)

Project base	Households recruited	Tonnes Greenhouse gas abatement per household per year
Local knowledge	611	5.02
Specialised knowledge	361	4.55
Strategic knowledge	768	8.01

* Individual and holistic knowledge were common to all.

- Are all the specialists taking full account of whole-of-community knowledge? Have they established connections to each other? Is their contribution taking precedence over and masking other knowledges? (specialised knowledge).
- Is there a whole-of-community strategic direction? Does this include the value orientations of citizens, specialists, government and industry? (organisational or strategic knowledge).
- Is there a clear focus on the transition to sustainability and the need for greenhouse gas abatement in this community? Does this focus allow for a synergy able to drive the generation of new knowledge and skills? (holistic knowledge).

The 22 communities were chosen from over 140 applications using the selection criteria:

- more than 50 households in size;
- household-focused rather than industry-focused;
- experienced in community-based social or environmental change programs;
- established groups willing to coordinate the Cool Communities project;
- showing potential for behaviour change in areas targeted by a project.

The national program provided support to each community through:

Individual identity: It offered a positive national 'brand', with recognition certificates given to participating households, community leaders and partners.

Local engagement: Participants could ask their own questions about greenhouse action through participatory workshops, theatre, audits, advisory services, demonstration homes and publications.

Specialised support: A panel of technical experts was available to advise all projects. Discounted goods and services were arranged.

Organisational structure: Program facilitators were employed in each state and territory to work with participating Cool Communities, with steering committees established in each community.

Holistic focus: The program provided each individual with the same focus within which they could act.

Each community intending to change was already an island within a larger interconnected system. The local manager of the Cool Communities program entered into the change negotiations from off the island. The government-based project managers could well have acted as missionaries and tried to convert the islanders, redirecting their knowledge to their own purpose. They could have acted as invaders, reassigning the islanders' knowledge resources for their own ends. In the event, they acted as beachcombers, accepting that they were there on sufferance from the community, and making use of any locally generated opportunity to advance their cause.

The effect of the different knowledge sources of the local change agents is recorded in figure 6.3. With such a technical goal (abatement of greenhouse gases), specialised advice might have been expected to be most effective in bringing about change. It was, however, the least effective. Increasing local community knowledge of its own processes led to slightly more gas reduction. The success rate was almost double from applying strategic or organisational knowledge.

Of Kelman's three proposed stages of adoption of social learning, there was little reward for islands and communities in stopping at compliance. Project support and rewards had been strategically related to advances in knowledge, values and skills. The next stage, identification with the project, was built into the negotiations from the start. The final stage, successful internalisation, was reached in 20 of the 22 communities that continued with the program (using their own internal resources) after the three-year support program finished.

The lessons from Cool Communities can be considered in any aspiring community linking social and environmental change. These 22 examples are only what could be, however. Any community seeking whole-of-community change would need to negotiate the conditions and monitor the outcomes for themselves. Greg Dening has the last word on guiding community change:

"Islands and beaches, of course, are everywhere – in a jungle clearing, in an urban ghetto, within a social class. Everywhere where space and action are contained by boundaries which screen comings and goings, there is an island and a beach. Every islander has to accept ideas from across a beach to

Figure 6.4 Pressure-State-Response framework with Potential for change added (d4-p4)

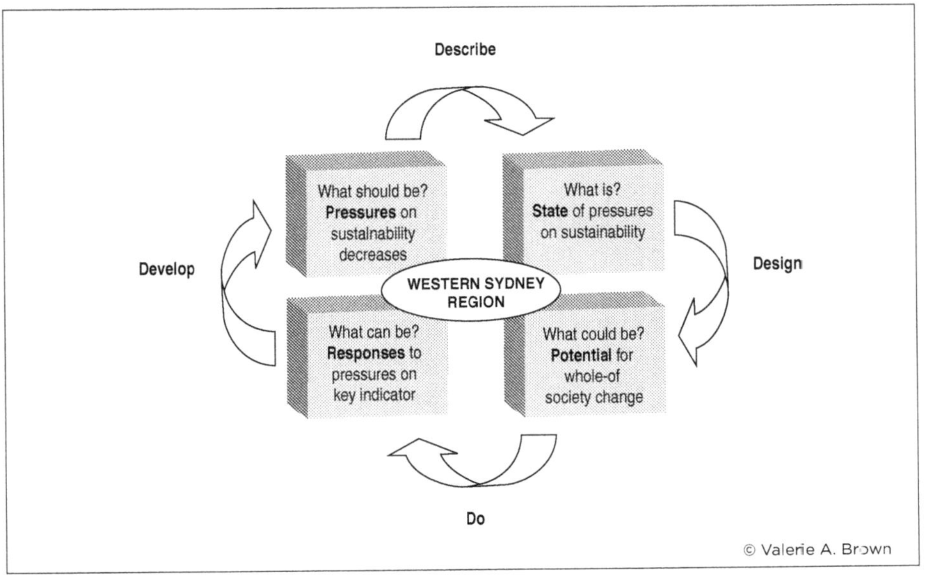

construct a new society. Across those beaches every intrusive artefact, material and cultural has to come ... On the land, behind the beach, life is lived with some fullness and some establishment. Crossing beaches is always dramatic, from either side." (Greg Dening *Islands and beaches* pp. 31–2)

What can be:
networking knowledge as a tool for change

Two further projects offer insights into a synthesis-based approach to whole-of-community change. Each followed the decisions-into-practice learning cycle, d4-p4. In the first project, whole-of-community sustainability monitoring, the knowledge cultures are treated separately as nested knowledges and the synthesis comes in a shared vision of the whole for a sustainable future. The second project, whole-of-community engagement in the sustainability of a vast river basin, is synergistic, based on continuous interaction among the knowledges. The aim in this project is to maximise collaboration, with tools that enable networking the knowledges.

Whole-of-environment monitoring

Communities trying to change how they live require continuous feedback to maintain a forward direction. For whole-of-community change, such feedback needs to be in terms that are accessible to all the knowledge cultures. An urban region that forms part of one of the world's spreading megalopolises sought to monitor its progress towards sustainability. The nine local government councils of the region, who together formed the Western Sydney Organisation of Councils (WSROC), decided to sponsor whole-of-community monitoring. The aim was to enable integrated decisions between the region's decision-makers.

There was a government requirement to apply the traditional state-of-the-environment monitoring framework developed by the Organisation for Economic Cooperative Development (OECD). This Pressure-State-Response (PSR) cycle had been in use in all industrialised countries since 1993. Its purpose is to document the pressures on the state of the environment from human activities, and match them to appropriate responses to those pressures. In the spirit of the claims in chapter 4, that any social learning process will follow the stages of the decisions-into-practice (d4-p4) cycle, government and community agreed to add a further dimension to the PSR cycle, that of 'Potential' (figure 6.4).

Changing the standard closed-cycle monitoring to a change-oriented spiral was only the beginning of the changes that were required. Eighteen months of team building and negotiation in a collaborative action research project was needed to establish the monitoring design. The research project team applied the methods of synoptic inquiry outlined in chapter 4. They worked in the spirit of critical loyalty; that is, strong mutual commitment between researchers and local government staff to whole-of-the-region monitoring. The result was a prototype to monitor sustainability. As we shall see, putting the design into long-term practice was another matter again.

The basis for the sustainability indicators was laid in 30 community vision workshops held in seven sites across the region between December 1999 and May 2000. The purpose was to identify community priorities and their matching community-based indicators for a sustainable healthy environment for the population of 1.5 million. The three-stage consultation process involved a pyramid of the nested knowledge cultures:

- a community-based futures visioning process, with 30 community workshops attended by more than 75 Western Sydney regional organisations and 900 participants, who together identified a preferred future for Western Sydney; followed by
- a government-based consultative workshop of 60 people, in which community advocates, nominated from each workshop group, worked with local and State government representatives to refine and find practicable measures for the indicators; and then
- a specialist-based forum for 30 people, whereby the nine council state-of-the-environment writers, 15 social and environmental scientists and six community delegates from the second workshop established measures, and designed tables, graphs and maps to convey the state of the eight environmental themes and the potential for achieving the community's preferred future to all the groups in Western Sydney.

Figure 6.5 WSROC Regional Community Vision, 2000

Once agreed on a common strategy, a specialist forum refined practical measures for the social and environmental goals of the change towards sustainability (figure 6.5).

The measures selected as indicators were 'headline indicators' in that they carried the main message the community wished to send to other decision-makers and the media. They provided symbols shared by the community, by which community members could recognise changes in the course of their everyday life (figure 6.5). They were also directly linked to council services so government lines of responsibility were clear.

An evaluation by Cesidio Parissi two years later found that the State government had recommended the report format to all the State's regions. Of the nine sponsoring councils, however, only one council was still following the process in full. Five were using a modification that still included community input. Three had reverted to the pre-existing specialist 'list of environmental facts' type of report.

Was this failure or success? On the one hand, there was now a framework for networked knowledge reporting, linked to a decisions-into-practice learning spiral that had not existed before. The framework had spread beyond the region and was being adapted to the needs of communities elsewhere. On the other hand, the paradox that lies behind combining community futures and larger social change proves to be alive and well. A social transformation tool can only work as well as both the implementing community and the external environment allow.

Whole-of-community engagement

A toolkit for whole-of-community engagement in guiding social change was the outcome of an action research study. Conducted by a human ecologist (this author), a social scientist (Heather Aslin) and a regional administrator (Alison Reid), the site of the research was the Murrumbidgee Irrigation Area (MIA), the largest food-producing region on the dry continent of Australia.

All available irrigation water in the area was being used. Global warming had contributed to a drop in rainfall over the entire catchment. And the potential for productivity for the region, tied to new overseas markets, was on the rise. The combination of these factors was placing unprecedented pressure on individuals, communities, expert advisors and governments alike.

The regional management system resembled the complexity of that of the Great Lakes of America. There were four sovereign states, 25 local councils and 12 integrated catchment management authorities. In this case, there was a population of only 300,000. The region's Community Advisory Committee held a wide-ranging consultation to establish some guiding values. They were:

- Courage: vision, leadership and taking difficult decisions
- Inclusiveness: trust, partnerships and ensuring full engagement
- Commitment: long-term view, stability and a non-partisan Basin perspective
- Respect: tolerance, honesty and acknowledging partners' realities
- Flexibility: reform, change and continuous improvement
- Mutuality: shared responsibility and accountability
- Practicability: long-term outcomes and viable solutions.

The study results confirmed that the compartmentalised knowledge hierarchy we reviewed in chapter 3 was alive and well in this large river basin. Further, each blamed the other knowledge cultures for the ongoing failure to come together in mutual cooperation. However, the values exercise revealed that each knowledge culture was able to share the principles held by the others. Combined, the knowledge cultures offered the versatility, reliability, strategic direction and holistic thinking that all sources confirmed that they needed. The challenge was to bring them all into a position to work collectively.

The outcome of the study was the publication of a toolkit based on the review of over 100 strategies for whole-of-community engagement. The toolkit was trialled with the same people who gave the original interviews, with their changes incorporated in the final version. A summary of the use of the toolkit in practice is in figure 6.6.

The experiences of community-based social change described in this chapter lead to three conclusions. First, there is a considerable range of tools on which to base the introduction of whole-of-community decision-making for social transformation. They must, however, be

Figure 6.6 Strategic tools for whole-of-community engagement (d4-p4)

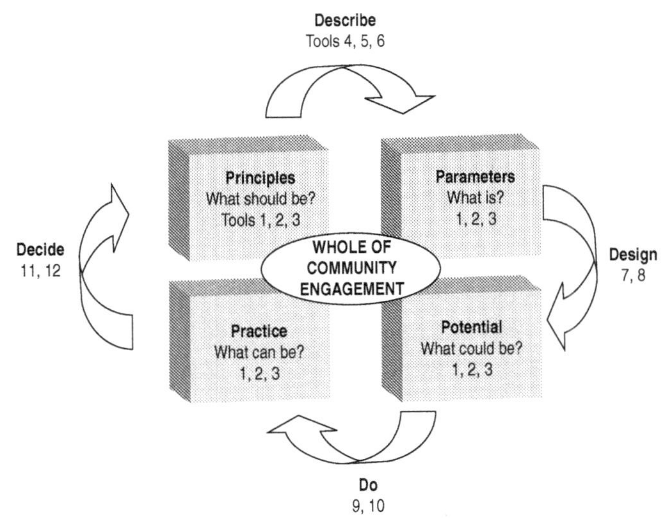

Type of strategic tool:

Generic tools:
1. General public involvement and participation
2. Negotiation and conflict resolution
3. Information, education and extension

Descriptive tools:
4. Rapid and participatory rural appraisal
5. Stakeholder analysis and social profiling
6. Survey and interview

Designing tools:
7. Planning and visioning
8. Teambuilding and leadership

Doing tools:
9. Participatory action research
10. Deliberative democracy

Developing tools:
11. Lobbying and campaigning
12. Participatory monitoring and evaluation

sensitively adapted to each particular community, whether large or small, they must engage all facets of the community, and they must cover all the learning stages.

Second, respect for local community knowledge and local community capacity to act is paramount. The changes won't even enter the community proper if they come only from the society outside. Going back to the theory contributions from Geertz and Putnam, an important link is the strategic social capital that links bonding (internal support) and bridging (external support) social capital. The tools listed in figure 6.6 can be used to forge all three. Third, commitment to change must have a holistic focus, a focus strongly held by action groups simultaneously at the local and the societal scales. Another way of putting this is to use Dening's metaphor of islands and beaches. If strangers land on the beach with a mission to change the social island formed by the community, a negotiation point needs to be recognised by both sides. Otherwise the result can be a destructive conflict, rather collective action towards a common cause.

There is a marked contrast between the difficulty of arriving at whole-of-community thinking in the 21st century with the ongoing admiration of the world for the Italian Renaissance cities of Florence, Siena, Urbino, Pisa and Rome. Floods of tourists from throughout the world visit their communities each summer, not only to admire the city architecture and look at the Renaissance painting and sculptures. Visitors wax lyrical about the ambience of the cities, the continuing sense of community, and the integrity between community and city design. If it could happen then, it can happen now.

In this chapter we have explored the use of symbols and open learning frameworks where the entry point to collective decision-making was a whole-of-community vision. In the next chapter, we move up a step through the nested knowledge cultures to consider how specialised knowledge can form the entry point to collective social change. We consider what it would take for specialists to not only go outside their own specialisation, but specialised knowledge as a whole, so as to include individual, community, organisational and holistic knowledge in a transdisciplinary synthesis.

7 Commonsense inquiry

"He must be a poor creature indeed who does not feel and know much more than he can give reason for. Hence the distinction between eloquence and wisdom, between ingenuity and common sense."

William Hazlitt Essay:
On genius and commonsense.

Consider the question, Why is that man being taken to be hanged? After a standard analytic inquiry, the answers might be: because he killed a man; because a jury found him guilty; because his country has the death penalty; because someone sold him a gun; or because nobody stopped him. The answer depends on the interests of the inquirer. A commonsense inquiry seeks a different set of answers: Was the sentence just? What were the social conditions? How could the killing have been prevented? In both analysis and synthesis, the answers have to be subjected to tests for truth. In a collective inquiry, what questions should we ask? How can we make sure the answers can be trusted?

What should be:
understanding the whole

It is little wonder that an era ushered in with Galileo Galilei's telescope and Newton's mathematics was named The Enlightenment and The Age of Reason. While the Renaissance revived belief in the human capacity to create new ideas and produce innovative technologies, this was always in the shadow of the temple and the palace. With the swing of Toulmin's pendulum, a fresh way of thinking began to ask questions generated by the Renaissance legacy. Leonardo da Vinci had led the way in his powerfully accurate observation of physical and social environments. But the idea that knowledge was a human construct, not handed down by a god or tribal chiefs, was a dangerous heresy to the established powers.

Two questions opened the door into the era of analysis, both about how we can come to know what we know. In the first half of the 16th century, Descartes the mathematician asked how we can know anything at all. He concluded that all we can really trust is 'I think, therefore I am'. In the second half of the same century, Montaigne the French essayist continually asked his readers 'What do I know?' He was questioning the right of authority to impose anything on anyone else other than what one had tested for oneself. From this point on, the rules for constructing new knowledge were doubt of anything that went before, and fresh testing by objective observations. The good scientist is therefore one willing to test all assumptions, to challenge all traditional opinion, to get closer to the truth. Science at its best is always aware of its limits, aware that knowledge is always growing, never absolute, and depends on evidence and reason.

The use of the scientific methods of observation and measurement doubled human life expectancy, led to instant global communications, the defeat of major epidemic diseases, high productivity in a global economy and humans living in space. It is said that specialist tools have conquered space and tamed wilderness. However, it must also be said that use of the same tools has led to the disruption of the Earth's life-support systems of water, air and energy; to whole populations and occupations displaced by technological change; to a gap between those who do and do not benefit from the new technologies; and the threat of atomic warfare.

The capacity to take systems apart needs to be matched by the ability to put them back together again. That requires a synthesis that includes the objectivity of Descartes' Enlightenment and the humanism of Leonardo's Renaissance. The most comprehensive of multidisciplinary teams is unable to advance human well being or environmental sustainability by the accuracy of their knowledge alone. For that, as argued throughout the previous chapters, specialist knowledge needs to be joined with individual commitment, community engagement, organisational efficiency, and a collective focus. Science is strengthened, not weakened, by joining in this collective understanding.

In other words, the objectivity of the enlightenment is a powerful tool for thought, but not the only one. Chapter 3 explored several ways of interpreting the world, each with its own tests for validity and truth. Chapter 5 described other intelligences, including the emotional, intuitive, and aesthetic. Personal experience and commonsense tells us that each way of thinking and each type of intelligence is always involved in the operation of scientific method, however much they have been suppressed. The practice of science is itself a belief system, privileging rationality and objectivity. Specialists develop a loyalty to their specialised community of practice. Nobel prize winners report that their leap in thinking came before, not after, collecting the evidence. Strong emotions can be released if the infallibility of science is questioned.

A common claim is that specialised knowledge is value free. Values (*what should be*) are ethical positions arising from a common humanity, inescapably connected to the emotions. Facts (*what is*) are conclusions about aspects of reality, based on specified sources of objective evidence, with the supposed exclusion of the emotions. The two are not independent of one another. The choice of subjects for and interpretation of the results of objective inquiry are both matters of values. On the other hand, values have no meaning unless they can be expressed in real situations.

The analytic era has given privilege to the pursuit of facts, to an extent satirised by Charles Dickens' character Mr Gradgrind, the successful 19th century business man and educator: "*Now, what I want is, Facts. Teach these boys and girls nothing but Facts. Facts alone are wanted in life. Plant nothing else, and root out everything else.*"

Specialised responses to wicked problems are more likely to echo Mr Gradgrind than to look to other modes of knowledge to assist in the resolution. The usual pattern of response among the disciplines is to do more of what they do so well already. That is, they increase the degree of analysis and add to the number of fellow-disciplines, rather than look for contributions from outside the specialisations. As an example, the field of linguistics has responded to its increased importance in this information age by dividing into anthropological, systemic, mathematical, area, text, historical and structural linguistics, and even more finely into ethno-, psycho- and socio- linguistics. Compare that focus on analysis, with Leonardo's combination of science, art and humanity in his drawing of a face in figure 7.1.

Figure 7.1 Leonardo da Vinci. *Light striking a profile*. **Florence, Uffiizi**

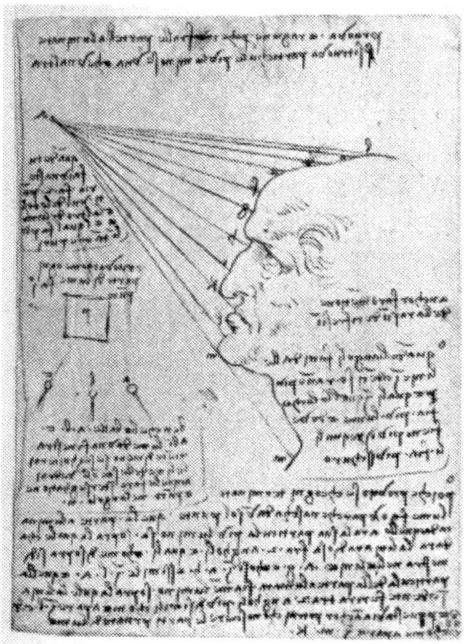

In addressing the complex nature of wicked problems, multidisciplinary teams are now the rule rather than the exception. Their dramatic successes are routinely attributed to the specialised contributions of the members of the team. There is rarely acknowledgement of the roles of the wide differences in the personal approaches of the individuals concerned, the loyalty of the community formed by the team, and the crucial importance of the style of management. This is illustrated in the saga of the Severe Acute Respiratory Syndrome (SARS), the first threatened pandemic since the 19th century risks of cholera and typhoid (box 7.1).

In considering the role of the specialised knowledge, the words specialist, scientist and expert are constantly being confused. Science tends to become the prototype for all specialisations, and expert to be synonymous with scientist. Single disciplines are assumed to be the model for all inter- and multi-disciplinary relationships. Clearer working definitions of these terms can be found in box 7.2.

In Chapter 4, we described how the knowledge cultures share Kolb's open learning framework in their construction of knowledge. They also share the framework for generating disciplinary knowledge first identified by Thomas Kuhn in *The structure of scientific revolutions*. Each mode of inquiry has its own particular method of investigation, type of question, test for truth and explanatory framework. Following Kuhn, in figure 7.2 the elements of specialised knowledge are matched against those of each of the other knowledge cultures. This allows both specialists and each of the other knowledge cultures to recognise the parallels between their constructions of knowledge. Together they form a sixth, collective construction of knowledge.

Each of the knowledge cultures has its own particular question, which it has

Figure 7.2 Five paradigms of knowledge (after Kuhn 1970). (Brown 2005)

	Contributors to collective decision-making				
	Individual	**Community**	**Specialised**	**Organisational**	**Holistic**
Dimension					
Content:	Personal lived experience	Mutual place-based experience	Academic disciplines, professions	Agendas, regulations, precedents	Symbols, metaphors, images
Inquiry:	Reflection	Dialogue	Observations	Cost/benefit	Imagination
Question	***Introspective***	***Social***	***Empirical***	***Strategic***	***Aesthetic***
Validation:	Self-reference	Local stories	Generalisation	Will it work?	Recognition
Explanatory frameworks: focus	Sense of identity	Community ethos	Cause-and-effect linkages	Strategic	Essence, plans
Role models:	Personal heroes	Eminent citizens	Nobel Prize winners	Powerful leaders	Writers, artists
Epistemology	Polanyi, Berlin	Geertz, Ife	Popper, Ravetz	Bordieu, Foucault	Hughes, Clark

become expert in answering. Although it has become cultural practice to only ask one type of question at a time, the questions can never, in actual practice, be separated. Leonardo's drawings in figures 7.1 and 7.4 appear to present only scientific observations. Yet each drawing is a combination of aesthetic flair, accurate observation, social communication (the drawings themselves), and a personal style.

Taken together, the knowledge cultures offer a set of the possible questions that can legitimately be asked in seeking to understand the whole. Philosophers long concerned with different constructions of knowledge confirm the validity of each question type. Examples of these thinkers are listed in the bottom line of figure 7.2, and in the bibliography. Scientific method asks empirical questions, ones that can be answered by detached observation of the physical world. On the other hand, social questions must be referred to the norms of the communities themselves. Strategic questions, the concern of organisations, are in essence ethical questions of who should benefit and how the benefit is to be achieved. Holistic questions are answered from the aesthetic satisfaction of a creative leap. Individuals implicitly ask all of these questions about anything they do. Socially, they have learnt to acknowledge only one type of question at a time, matched to a specific context.

What all the questioners share is the claim to be using commonsense. Mathematician Rene Descartes proposed a

Box 7.1
COLLECTIVE KNOWLEDGE OF A WICKED PROBLEM: SARS (SEVERE ACUTE RESPIRATORY SYNDROME)

The wicked problem

In January 2003, an infectious disease erupted into a world that thought it had the capacity to control such diseases. In January 2003, Severe Acute Respiratory Syndrome (SARS) was a new and threatening public health problem crossing regional boundaries and unresponsive to the usual controls. SARS emerged in a rural province of China, and WHO rapidly identified the high infection rate and the high case fatality rate of 9.6% as of immense concern. Contrary to any expectations, infection rates were higher in hospitals and among hospital staff than in the general population.

Specialist responses

SARS reached 27 countries in 3 months, due to high urban density and rapid transport around the globe, especially by air. Containment was by known infectious disease controls, once the virus, source and transmission were identified. Cultural investigations eventually found that SARS arose, apparently spontaneously, through a traditional human/environment interaction: Chinese fondness for the meat of wild civet cats. An ecologist then identified a genetic mutation from a bat virus, transferred through sharing the same marketplace. Between the World Health Organisation and the Internet, progress of the disease could be followed by media, community and politicians, medical scientists. and clinicians. Their collaboration with each other and with governments and social service orgnaisations led to this disease with the potential to decimate the global population being contained within 6 months.

Individual responses

Both the spread of SARS and its containment were influenced by the actions of particular individuals. Two medical practitioners in China who developed SARS symptoms fled, one to Hongkong and one to Saigon. Both eventually died of the disease. In Hongkong, the individual booked into a hotel, from where he infected a lift-full of people. In Saigon the individual isolated himself, and refused to let other health practitioners treat him, in order to contain the infection. Hongkong had over 1700 cases, Vietnam 63. In spite of Chinese government censorship, a Beijing surgeon released hospital figures to Western media, allowing WHO to identify transmission points. The Chinese doctor was removed from his post. Two research scientists did not take the advised precautions, contracted SARS and in traveling to Singapore and Taiwan led to extreme isolation measures and so major economic disruption.

Community responses

It took a mere three months for SARS to move from a few instances of atypical pneumonia to a potential global threat, The worldwide information network allowed almost instantaneous dissemination – China was initially the exception. WHO broadcast daily bulletins on the Internet in clear lay persons' languages to keep people informed worldwide. Nevertheless, there

Box 7.1 continued

was community panic, in China, Asia and Western countries connected by air, there was a widespread drop in travel, and the wearing of face masks described as useless by WHO. New communities of researchers and of clinicians formed to share methods and findings. Villages in China replaced their village walls and controlled access for the first time in centuries. Private industry and government agencies not necessarily friendly to each other collaborated freely in local communities around the world.

Organisational responses
WHO coordinated an extraordinary global surveillance and research information system. Transdisciplinary teams went to every stricken country and region. Research laboratories who usually compete for drug company and government funds shared their results freely. Clinicians shared anecdotes without waiting for their usual level of certainty. Village workers freely shared their experience with government agencies. If such teams had been in place before the event, monitoring prospective risks and responses, SARS may have been prevented altogether, or at least stopped at source. Here is a precedent for the future.

Collective minds
Specialists, individual, community and organisational responses a. worked to a shared holistic focus and b. collaborated across external and internal divisions. The outcome was unprecedented success in terms of time and effectiveness in controlling a global threat.

Box 7.2
EXPERT, GENERALIST, SPECIALIST AND SCIENTIST: A GLOSSARY*

Expert: a person with a high level of skill in any field, trade or profession e.g. microsurgery, skiing or car repairs.

Generalist: a person who is equipped to access a range of ways of knowing, e.g. human ecologist, systems analyst, administrator.

Synthesist: a person who has skills in bringing together different ways of knowing, e.g. theatre producer, orchestra conductor, systems designer, strategic planner.

Specialist: a member of a particular community of professional practice, with its own disciplined mode of practice and methods of inquiry e.g. physicist, philosopher, historian, economist.

Scientist: a professional practitioner whose mode of practice is objective and rational: e.g. oncologist, ecologist, astronomer, sociologist, behavioural psychologist.

Monodisciplinarity: membership of a single disciplinary practice, the standard unit of universities.

Multidisciplinarity: collaboration between members of different disciplinary practices.

Transdisciplinarity: collaboration including, and going beyond, the different disciplinary practices, to include all the knowledge cultures.

Profession: Commitment to a particular form of construction and application of knowledge, its ethics and its institutions.

(*Terms often used interchangeably. These are the working definitions being used in this book)

commonsense that rigorously excludes the subjectivity of the observer. Anthropologist Clifford Geertz restored the subjective in his definition of commonsense, previously quoted in chapter 3, as *"citizens not just using their eyes and ears, but using them collectively, judiciously and reflectively to understand their own locality"*. Asking the full set of questions allows objectivity and subjectivity, facts and values, to be re-united.

The following set of commonsense questions has been canvassed in detail in this author's 1981 *From sickness to health: an altered focus for health care research*. They have been applied in practice in her 1985 *Towards an epidemiology of health: A basis for planning community health programs*. Here we look for the practical links between the specialised and the rest of the knowledge cultures, leading to collective practice.

Six commonsense questions

In everyday decision-making, the rationality of specialised knowledge needs to be linked to the subjective experiences of individuals, the social purpose of a community, and the pragmatism of an organisation. The full range of specialist questions can be incorporated within the possible types of commonsense question, that follow (see fig 7.5):

Empirical questions:	What observations can we make?
Social questions:	What do people here hold in common?
Aesthetic questions:	What is experienced here as beautiful or ugly?
Ethical questions:	What is accepted here as being good or evil?
Introspective questions:	What do I think about this?
Reflective questions:	What does all the evidence add up to?

In a synthesis, all the commonsense questions need to be answered, just as all knowledge cultures need to be involved. Not all cultures are equally able to answer every question. Just as specialised knowledge gives more weight to empirical questions, so does the individual to introspective questions, the community to social questions, the organisation to ethical questions, and the holist to aesthetic questions. In everyday language, bringing these together on any given theme is the equivalent of combining individual and local knowledge, science, politics and art.

The nested and networked set of knowledge cultures that we discussed in chapter 3 can be related to the use of the six commonsense questions in figure 7.3. Take a practical example of the need to ask the full set of questions. Consider the case of open-heart surgery. Anaesthetist, surgeon, community worker, hospital administrator, nurses, family and friends all make essential contributions to the patient's well being and the success of the operation. In standard practice, the success is attributed only to the surgeon. All

Figure 7.3 Construct mapping of the 6 elements of collective knowledge

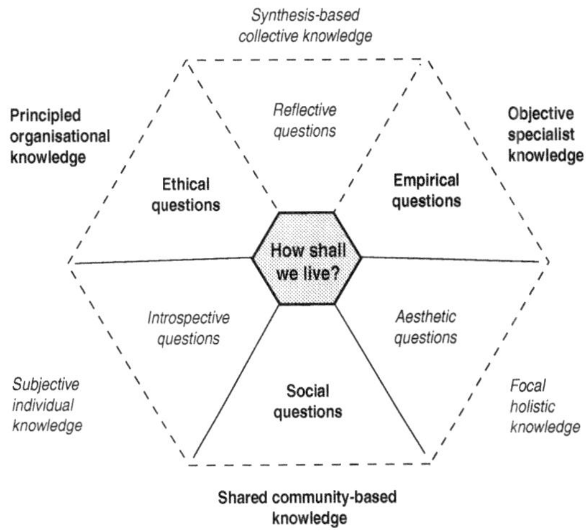

© Valerie A. Brown

six questions are of importance to the patient, but will be answered separately by the different players.

Thus the surgeon's outcome measure will be a textbook recovery and a return home in five days. The anaesthetist will calculate the patient's sedation according to weight, age and health status but usually without knowledge of their level of stress or individual response to pain. The community worker may be aware of difficult living conditions for the patient, but is unlikely to be consulted by the hospital team. Family and friends will know whether they will be available for care at home. The individual patient will have different concerns about different aspects of the procedure and aftercare. The degree of specialisation works brilliantly for the success of the physical event, the open-heart surgery. If the evidence from the different sources is not reconciled, however, there is a likelihood of a post-operative disaster.

It is important to remember, as we noted in chapter 1, that synthesis is not a fixed outcome, but a learning process. The comparisons in the surgery example above are not meant to claim any one form is preferable to the other, but to point out that there are many pathways for initiating collective knowledge, depending on the purpose. The key point of moving to synthesis is the necessity of asking all six of the types of commonsense question. This requires strategies capable of bringing the knowledge cultures and the commonsense questions together. These are reviewed under 'What could be' and 'What can be' sections below.

First, however, it is necessary to consider aspects of specialised knowledge that make transdisciplinarity difficult to achieve.

What is: thinking as hedgehogs, foxes or hounds

Figure 7.4 Leonardo da Vinci *Sketch of a perpetual motion machine* London, Anne Ronan

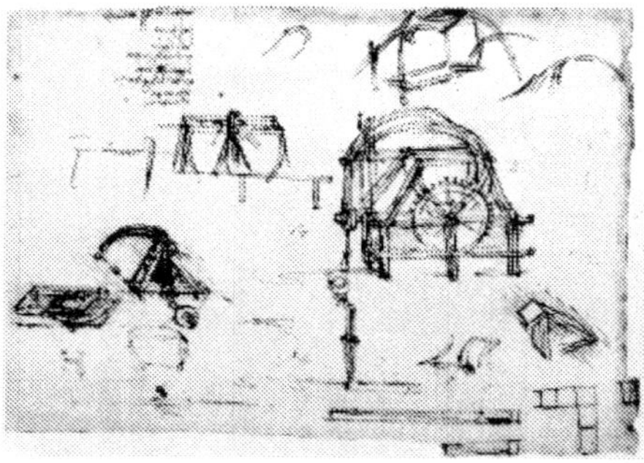

Kuhn's description of the normal scientist in *The structure of scientific revolutions* is as someone who works within the security of an existing framework. In studies of the partitioning of individual thinking styles listed in chapter 5, in particular the Myers-Briggs analysis, people working in science have been found to be introverted rather than extroverted, thinking rather than feeling, and judgmental rather than perceptive. These are not characteristics that lead to working readily with collaborators outside their own field. However, the American sociologist and philosopher of knowledge, Jerry Ravetz, has identified open-endedness and creativity as increasingly common practice in normal science. For this trend he has generated the phrase 'post-normal science'.

In chapter 5 we traced the use of the hedgehog and the fox metaphors for synthesis and analysis through an early Greek poet to the Renaissance, and on to Isaiah Berlin's review of classic writers. Neither the image of hedgehog (as someone committed to illuminating a central theme) nor of fox (as a person who picks up useful ideas as they go along) seemed to adequately explain the modern scientist. We therefore introduced the metaphor of specialists as packs of hounds, working in their disciplines, finding a trail and pursuing it to the end.

The fragmentation of evidence is reinforced by the process of specialisation. Box 7.3 contains descriptions of the same physical environment from an ecologist born there, a social scientist hoping to improve mental health there and a natural scientist planning to manage the environment there. The writers are using reflective, social and empirical evidence, and their style of presentation matches the metaphors of hedgehog, fox and hounds. Each of the three is implicitly answering all six types of question – aesthetic, ethical, introspective, social, empirical and reflective – but only one type is made explicit in each case. No one of the writers recognises the role of others. Yet any effective action towards a sustainable future for the region would need the combined expertise of them all. Compare this again with da Vinci's capacity to use the facts as they are known at the time, imagination and artistic expression in his drawing of a perpetual motion machine (figure 7.4).

The poet Judith Wright has something to say about scientific objectivity taken to extremes, in her poem

Box 7.3

THE MALLEE COUNTRY: MULTIPLE REALITIES

1: Ecologist and resident: introspective

I spent my early childhood in the Mallee Country. There is a funny feeling about it when you first go back there. It's like a sense of belonging, a familiarity, a wonder at its beauty and uniqueness. Its beauty is in the colours of the red sandy soil, green tips of the mallee, land forms, dunes, wildlife, birds. During drought times it is extra powerful. Beauty is associated with that joy, the whole cycle of dust storms and hot dry weather. The forces of nature are evident and of course more powerful than our actions on the ground.

www.nexus.edu.au/TeachStud/arera/research/hard.HTM [accessed 12.12.04]

2. Social researcher: normative

The Southern Mallee region (is) located in Victoria's North West and covers an area of approximately 59,150 sq km. The population of 280,000 is predominantly Australian-born with indigenous communities in the north and centre of the region. The region is the most disadvantaged in the state. Using a multidisciplinary approach (psychology, psychiatry, sociology, demography) we have developed a framework for the way in which individual psychological factors might interact with geographical, social, cultural and economic aspects of place.

www.dhs.vic.gov.au/careers/pages/loddon.htm [accessed 12.12.04]

3: Environmental manager: empirical

The Rose Mallee (Eucalyptus rhodantha) is a low spreading mallee that grows up to four metres high with smooth greyish brown stems and bluish green branches. The flowers are pollinated by several species of birds and small mammals including the White-fronted Honeyeater (Phildonyris albifrons) and the Honey Possum (Tarsipes rostratus). Threats to the Rose Mallee include land clearing and the drift from herbicides and pesticides sprays, and predation by feral rabbits. Listing under the Commonwealth's Endangered Species Protection Act 1992 requires that a recovery plan be prepared for the Rose Mallee.

www.affa.gov.au/ministers/ anderson/releases/97nhtlaunch.html [accessed 12.12.04]

Eve to Her Daughters (box 7.4). The poem echoes the many critiques of specialised knowledge as perpetuating a Western, middle-class, male, patriarchal perspective on the world. Vandana Shiva's *Staying alive: women, ecology and development* documents in detail how scientific communities marginalise and reject knowledge constructed by non-Western cultures, local communities, and women. Reading Wright's poem encapsulates the evidence that belief in objectivity is a subjective human practice, with important human consequences.

More recently, there have been pleas from within science for the reunion of the now divided ways of thinking. Notable examples are Edmund Wilson's *Consilience* (conscience plus resilience) and Stephen Jay Gould's *The Hedgehog, the Fox and the Magister's Pox* (the generalist, the specialist and the strategist). Gould, in his final book, criticises Wilson for continuing to privilege the objectivity of science

> **Box 7.4**
> **EVE TO HER DAUGHTERS**
>
> So he set to work.
> the earth must be made a new Eden
> with central heating, domesticated
> animals, mechanical harvesters, combustion engines, escalators, refrigerators,
> and modern means of communication
> and multiplied opportunities for safe
> investment
> and higher education for Abel and Cain
> and the rest of the family.
>
> In the process he had to unravel everything,
> because he believed that the mechanism
> was the whole secret - he was always
> mechanical-minded.
> He got to the very inside
> of the whole machine
> exclaiming as he went,
> So this is how it works!
> And now that I know how it works,
> why I must have invented it.
> As for God and the Other, they cannot
> be demonstrated,
> and what cannot be demonstrated
> doesn't exist.
>
> Yes, he got to the centre
> Where nothing at all can be demonstrated.
> And clearly he doesn't exist; but he
> refuses
> to accept the conclusion.
> You see, he was always an egotist.
>
> Source: Judith Wright 1976, Eve to Her Daughters in A Human Pattern p134

over the inspiration of the arts. Gould invokes the inspiration and joy arising from synthesis as "*the realisation of our highest mental and moral capabilities through a consilience of equal regard*".

What could be:
transdisciplinary inquiry

Members of all the areas of specialisation, including the scientific community, have several options for joining with other knowledge cultures. The following avenues for transdisciplinary partnerships are not mutually exclusive, nor do they reduce science's own effectiveness:

- **Miracle science:** continue with mono-, or inter-, or multi-disciplinary discoveries, and leaving other knowledge cultures to apply them;
- **Post-normal science:** develop systemic thinking, with a primary interest in patterns of relationships within science;
- **Citizen science:** support the capacity for individuals and communities to apply scientific methods for themselves;
- **Strategic science:** interpret the social and economic implications of scientific findings for political purposes; and
- **Sustainability science:** form alliances with the other knowledge cultures, for the shared purpose of re-establishing a sustainable humane environment.

The option of sustainability science is being promoted in research organisations around the world. An example is the United Research Councils of the United States' *Our common journey*. It is the option most explicitly directed towards a culture of collective knowledge. However, in practice it remains firmly based in the specialised knowledge culture and restricted to empirical science. Within the science community, the potential for transdisciplinary enterprise is almost always a matter of recruiting a multi-science team. Examples include the Harvard and Cornell sustainability science programs, and most university courses on sustainable development.

Within the specialist tradition, the closest to the full synoptic reach of transdisciplinarity falls under the heading of strategic science. The main avenue for delivering the whole is the development of critical systems thinking. C. West Churchman, one of the founding fathers of the fields of operational research, management science and the systems approach, is also one of transdisciplinarity's major critics. In *The Systems Approach and its Enemies*, he urges these fields to remain faithful to the synthesis-based, self-reflective, transdisciplinary and ethical spirit that he embraced throughout his work.

Returning to the use of the full set of commonsense questions, this can be applied to an earlier example of a wicked problem, the health effects of Lead in air. One example of a collective response based on the set of commonsense questions is the removal of the industrial by-product, Lead, from the air that people must breathe. We have described the history of strategic responses to that issue in chapter 3. In figure 7.5 we match the commonsense questions to the appropriate disciplines for finding the answers. We do this in the context of collective action on Lead from within public health.

We take up again the story told in box 3.2, about the knowledge cultures' combined influence on actions on the social, economic and environmental impacts of Lead. The outcome of the integrated inquiry was a fresh policy of prevention of harm, lowering the permissible child lead levels, removing lead from petrol, and setting up a fresh study of the implementation of the policy and its effect on the sections of the Australian population at high risk. The study was one of the first transdisciplinary inquiries funded by the National Health and Medical Research Council.

The study, *Reducing Lead exposure in Australia: risk assessment and analysis of economic, social and environmental Impacts* was undertaken in 1993, by a multi-disciplinary team of Deni Greene, environmental scientist, Mike Berry, urban planner, and Jan Garrard, social scientist. Their consultations with the community, the Lead industry, the car manufacturers and politicians supply the following profile of answers to the commonsense questions:

Empirical questions about Lead levels: Questions about physical conditions and events-- chemistry of petroleum; decilitres of blood; micrograms of Lead; IQ measures; behaviours; costs; taxes. Answers are validated against calibrated standards and scales, many set by the World Health Organisation.

Answers: *Children under four absorb Lead in their bloodstream at eight times that of adults. When further epidemiological studies were conducted on children aged one to*

Figure 7.5 Commonsense questions and the transdisciplinary Lead study

Type of knowledge	Matching question	Source of answers
Empirical What and where?	Questions of people and place What is it made of? What resources does it use? Where is it? What is its social, physical, and ecological state?	Chemistry Demography Environmental sciences Community records
Social Who?	Questions of meaning What are the rules, and roles? What are the symbols, language? How do they distribute resources? How do they allocate power?	Community stories Sociology Economics Political science
Ethical Why?	Questions of purpose What is the purpose of this activity? What are its ideals? What are the reference points for right and wrong? Good and evil?	Philosophy Ethical principles Cultural history Spiritual commitment
Aesthetic Which?	Questions of feeling What are the preferred styles, forms, standards of address, behaviour? Principles of design?	Design, aesthetics Personal feelings Culturally accepted forms
Introspective Who am I?	Questions of identity What does all this mean to me? For someone of my age, income, education, family, experiences, what does the future mean?	Personal goals self-awareness Identity, experience Hopes, fears, expectations
Reflexive What shall we do?	Questions of a preferred future What does it all add up to? What actions does a community take?	A collective understanding of principles, parameters, possibilities and practice.

Basis for analysis: Greene, D., M. Berry, and J. Garrard. *Reducing Lead Exposure in Australia: Risk Assessment and Analysis of Economic, Social and Environmental Impacts.* Canberra: National Health and Medical Research Council, 1993.

five, the level of unacceptable harm was estimated to be 0 micrograms per decilitre.

Social questions about Lead levels: Questions about the discourses around the topic of Lead. These included I-You, I-Thou and I-It relationship among the actors on Lead; questions of intellectual development; paternal and maternal acceptance of responsibilities; level of acceptance of industrial risk; economic influences and outcomes, patterns of public decision-making. Answers are validated by confirming the presence of agreed meanings and shared interpretations.

Answers: *The initial public reaction in Lead-contaminated areas was, in social terms, to blame the victim (i.e. to place the burden of preventive action on the overworked parents of the low-income families in the high-risk areas). Mothers were instructed to be more thorough in cleaning their houses. After citizen advocacy, the level of social awareness rose, and active citizen lobbying of industry and government began.*

Ethical questions about Lead levels: Obligations of society to children; obligations of parents to children; obligations of society to low-income groups; professional ethics; consumer rights and responsibilities; organisational as opposed to individual rights; and the question of whether the environment has rights. Answers are validated by reference to ethical systems that determine right and wrong.

Answers: *The public health services thought it legitimate (ethical) to continue to use the higher risk level of 20 micrograms because to lower it would unnecessarily alarm the public, and reduce the economic viability of industries providing employment. When risks to the intellectual development of children became public knowledge, it was the community that found the risks to children ethically intolerable.*

Aesthetic questions about Lead levels: Questions about aesthetic choices and feelings – the invisibility of Lead particles; the smooth functioning of automobile engines; the desire to renovate inner-city dwellings that release Lead-laden dust. Answers are validated by reference to standards of feeling, form and style.

Answers: *Lead in air, water or soil has no effect on taste, smell or appearance, so it is not guarded against in the same way as faeces or decaying vegetable matter. The use of Lead is an optional additive to ensure better-running engines, an aesthetic and economic priority. Changes in tastes lead to renovation of older buildings, with their Lead hazard in the accumulated dust.*

Introspective questions about Lead levels: Questions about individual experience – the range of each individual's freedom to choose; individual experience of Lead contamination according to age, socioeconomic class, family relationships, geographic and cultural setting and life events. Answers are validated by their relevance to and integrity for the individual observer.

Answers: *The experience of Lead contamination was for a decade sharply divided between three sets of interests - the self-interest of health organisations, the ill-health of those with Lead contamination; and the lack of information on the issue for the population at large. In late 1980, the health consumer lobby decided to make environmental risks more transparent to the community.*

Reflexive questions about Lead levels: Questions about acting on the evidence — the evidence is evaluated ethically (the interested parties differ); socially (the Departments of Environment and Primary Industry); empirically (Public Health general research and local Lead impact studies); aesthetically (appeals are made to standards of 'what is appropriate' - they differ); and introspectively (against the experience of the Lead-affected).

Answers: *In the transdisciplinary study, 12 principal Lead abatement measures were identified, their social impact evaluated and subject to a cost-benefit analysis. Economic benefits rose with every drop in the reduction of Lead.*

The long-term outcome? A review was conducted ten years later by Deni Greene, the lead researcher in the transdisciplinary study. It found that the average Lead level of the nation's children had fallen to safe levels. Vulnerable sub-groups, such as children in Lead-producing areas and industrial workers, were still at high risk. Regulatory standards are monitored in random sites across the community, and not near point sources of Lead, so the results give neither practical warning nor result in prosecutions. Industry controls are largely self-policed. The government in power changed the monitoring committee to include industry only.

The moral? Collective decision-making enables the arrival at the 'right' decisions; that is, policies that serve both the individual and public good. Those 'right' decisions need to be followed by the 'right' conditions for action. Policy and practice would be better developed if the conditions favour collective decision-making,

In the next section, the set of common-sense questions are applied to a project intended to generate collective change throughout a profession. This was the preparation of a text-book intended to re-orient public health from local to global risks, from treatment to prevention, and from specialisation to transdisciplinarity.

What can be:
changing professions

Public health is one of the specialised fields most in need of collective decision-making. Health practitioners, engineers, town planners and public education made it possible to overcome the epidemic diseases emerging from the overcrowded cities of the 19th century. Ironically, one outcome of public health's own success is the lowering of mortality rates and so increasing the pressure of population growth on global life-support systems.

This is not the first time human activities have threatened human survival on the planet. In *The Future-Eaters*, mammalian archaeologist Tim Flannery described how humans' mental and technical skills have driven them to overrun the capacity of their environment right through human history and prehistory. The resulting crash in the available resources leads to fresh inventiveness which evolves into a new form of alliance between humans and the environment: a learning spiral. The difference in the 21st century is that the future-eating of today involves not just the resources of localised patches, but all the resources of the planet.

Ever since the 19th-century radical social change of the hygiene revolution, the direction of the public health profession has been to increasing specialisation on the one hand, and administration on the other. Public health in the 21st century is diversified into numerous specialisations, including environmental health, legislation, education, lifestyle change, toxicology, occupational health, nutrition, and community development. A study of Australian environmental health practice led by this author, *Grass roots and common ground: guidelines for community-based environmental health action*, found that professional practice has been directed towards administration of existing legislation, rather than radical action on the emerging global issues.

As a contribution towards addressing the public health outcomes from global health impacts, the national government funded the preparation of a transdisciplinary teaching-learning textbook.

That text has since been published as *Sustainability and health: supporting global ecological integrity in public health* by Valerie Brown, John Grootjans, Jan Ritchie, Mardi Townsend and Glenda Verrinder. The initiative was essentially transdisciplinary and collective in itself, in that it involved a project management team from five different fields in five different universities, and 10 international experts and 30 colleagues from individual, community, specialised, organisational and holistic approaches to health. The book was tested in eight trial courses. Dialogue between those many interests was established over a year and a half, through phone conferences, email discussion groups and an open access website.

The development process for the textbook involved all participants in collegiate use of the decisions-into-practice learning spiral. The process was synoptic, in that every step of the process involved all the key interest groups on an equal footing. It was synergistic, generating new ideas at each of the steps of the experiential learning and decisions-into-practice cycle (figure 7.7).

The holistic 'flavour' of the resulting text did not issue from any one author, nor from the editorial panel. It was the outcome of a collective process. The mutual learning strategy was designed by a broadly-based management committee. Internationally-respected experts working on the connections between sustainability and health came from research, social change, environmental monitoring, indigenous health, sustainable communities, global integrity, administration and the arts. The experts agreed that:

- sustainability in public health is about transformational change in environmental governance; it is not about revolution, but nor is it about 'business as usual';
- sustainability in public health requires local community knowledge and organisational strategic knowledge of the global environment as well as the specialised knowledge of the academic disciplines; and
- sustainability and health teaching programs will need to deal with the future as well as the present, and about what to preserve as well as what to change.

At the first stage of the learning cycle, it was agreed that the principles for sustainability were those of the Earth Charter (box 7.5). A logo was developed to represent the essence of the project and act as a shared symbol linking the collaborators (figure 7.6). Current parameters of global sustainability and health were established through a website discussion among all the potential users of the text. The potential for change was maximised during a three-day writing workshop, with each of five groups made up of six members, one from each of the knowledge cultures. This brought into play the full portfolio of tools for collective thinking described in chapter 4. As the change framework developed, participants met in their original knowledge cultures to evaluate the feasibility of its introduction into existing programs.

Box 7.5

KEY PRINCIPLES OF THE EARTH CHARTER*

1. Respect Earth and life in all its diversity
2. Care for the community of life with understanding, compassion and love
3. Build democratic societies that are just, participatory, sustainable and peaceful
4. Secure Earth's bounty and beauty for present and future generations

*website: http://ww.earthcharter.org (accessed 20.8.05)

Figure 7.6 Logo of Sustainability and Health Project

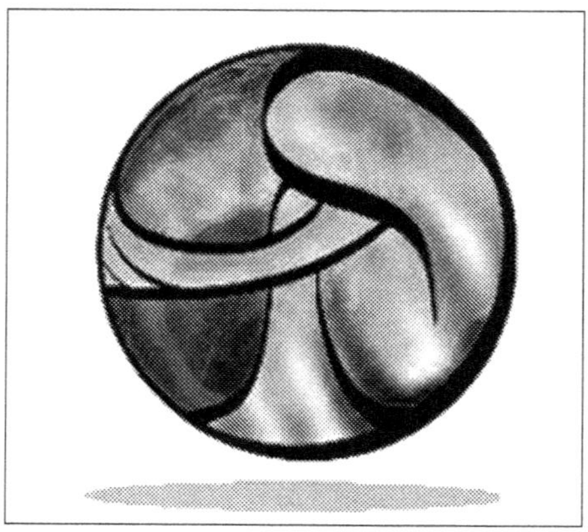

The result is a text that is written for the practitioner, translating technical terms into Standard English wherever possible. Each chapter title refers to an

Figure 7.7 Stages of Sustainability and Health change processes (d4-p4)

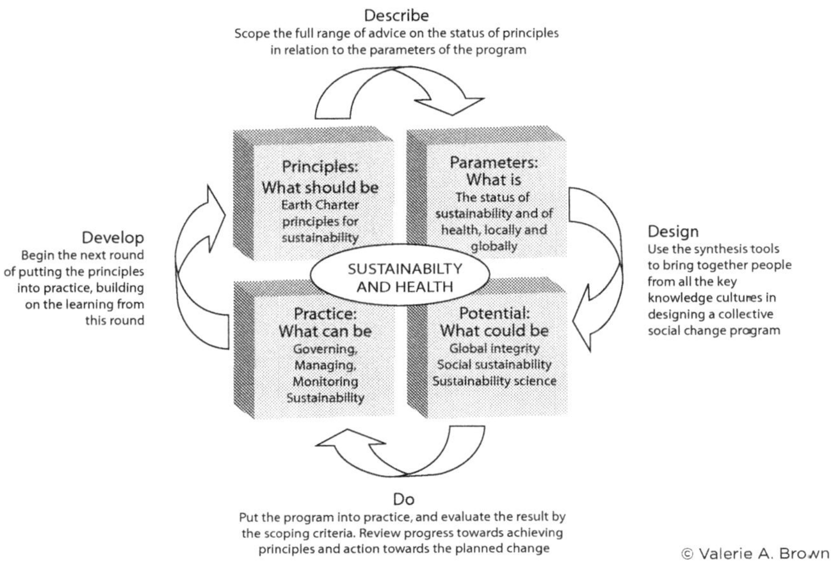

action, a 'doing' word: Living, Listening, Grounding, Knowing, Scoping, Acting, Innovating and Managing. Each chapter responds to one commonsense question in particular, but each chapter draws on all the knowledge cultures to answer all the questions. Throughout, the text attempts to meet specialist standards for referencing the evidence, community standards for understanding the process, and organisational standards for effectiveness of the process.

An independent consultant facilitated the three days of the writing workshop, ensuring that all participants followed the rules of dialogue and open learning described in chapter 4. The workshop evaluation confirmed the energy generated from the diversity of the participants, amid the confusion of some of the participants at what was for them a new experience. Each group was given the responsibility for two chapters of the book. Groups gave critiques of other groups' chapters, using the principle of critical loyalty. Amazingly, the chapters were largely completed by the end of the three days. The text is now in operation in Sustainability and Health courses world-wide.

During the workshop, participants regrouped at regular intervals into their original knowledge cultures. This grounded the innovative work of the synoptic groups in the practical demands of existing organisational structures. Evaluation of the professional change process and responses to the resulting textbook led to the conclusion that specialists wishing to support transformational change in their profession have four options. They can:

- commit themselves to **personal and professional change**, becoming a transdisciplinary service provider and scholar;
- contribute to the further skill development of their **specialised community** by introducing skills of synthesis as well as analysis;
- maintain their traditional role as **expert practitioner or researcher**, contributing their own case study examples and investigative skills to the synthesis;
- introduce transdisciplinary thinking as a **knowledge broker** within their organisations, and between their organisation and the other decision-making sectors; and/or
- become an **advocate** for the development of an underlying synoptic ethic, within which all the knowledge cultures can work collectively.

In choosing any of these options, specialists can make a leap forward from Descartes' rational and reductionist path that has served their world so well for so long, but which has created a monoculture often closed to new alliances. Any of the options would meet Descartes' criterion of 'rightly applying the mind' and contribute towards a more sustainable future.

The exercise in box 7.6 offers the reader the opportunity to develop his or her own integrated transdisciplinary inquiry. The next chapter examines the role of organisational knowledge in contributing to synthesis of knowledge and of action. We find that organisational knowledge provides the power base for other knowledge cultures to put their ideas into action. We also find that the other knowledge cultures are essential for providing organisations with the evidence, the focus and the commitment needed to meet organisational goals.

Giving the last word to a poet:

"I care not if you bridge the seas,
Or ride secure the cruel skies,
Or build consummate palaces
Of metal or of masonry.
But have you wine and music still,
And statues and a bright-eyed love,
And foolish thoughts of good or ill,
And prayers to them that sit above?"

J. E. Flecker 1900
To a Poet *One Thousand Years Hence* p7

Box 7.6

TESTING: IN SEARCH OF COMMONSENSE ANSWERS (TOWARDS TRANSDISCIPLINARY SOLUTIONS)

Consider an issue of importance to you, your occupation or your community.

Take the following questions:

Empirical questions:	*What observations can we make?*
Social questions	*What do people here hold in common?*
Aesthetic questions:	*What is experienced here as beautiful or ugly?*
Ethical questions:	*What is accepted here as being good or evil?*
Introspective questions:	*What is my own experience of this?*
Reflective questions:	*What does all the evidence add up to?*

Where would you find the answers to

A. Your ideal of what *ought to be?*

B. The collective evidence on *what is?*

C. The range of ideas on *what could be?*

D. *What can be*, right now, if we act collectively?

Put together a portfolio of where you need to look for the answers.
You now have a full profile for a transdisciplinary study of the topic or issue.

If you are planning a transdisciplinary research study, go to section 3 of chapter 4 and apply the questions in the scoping and conclusion stages.

8 Integrative management

"From space, we see a small and fragile ball dominated not by human activity and edifice, but by a pattern of clouds, oceans, greenery and soils. Humanity's inability to fit its doings into that pattern is changing planetary systems, fundamentally. Many such changes are accompanied by life-threatening hazards. This new reality, from which there is no escape, must be recognized – and managed."

Brundtland Report, World Commission on Environment and Development 1987

'Being democratic' is the mantra of 21st century management – used by individuals, communities, organisations and government to describe good governance. Taken literally, democracy means 'let the common people rule'. The term loses its meaning, however, when used to justify the activities of dictators and demagogues, and authoritarian and hierarchical organisations. The original ideal of democracy was for open government in which all people were equal, and government was for the good of all. Collective belief in these ideals seems to have lapsed at just the time in history when the scope of good management reaches the entire planet. A new confidence in holocentric management points the way towards a collective future.

What should be:

principled holocentric management systems

During the 20th century, governments and corporations became more and more alike. Government agencies now increasingly take on the form of a corporation. They have products, consumers, performance indicators and cost-benefit analyses. Social service agencies have adopted the business model, or been required to do so by their government funding agencies. At the same time, corporations have taken on many of the powers of the state. They set wage levels, determine working conditions and control the parameters for the development of whole suburbs and regions. This convergence between government, social service and industry management structures needs to be taken into account in examining how organisational knowledge can contribute to collective decisions.

In trying to cope with a complex world, the power of organisational knowledge has come to shape our society and determine the course of our social institutions of education, law, finance, religion and politics. In principle an organisation takes account of the individuals that work within it, the community in which it is set, and the expert advice it receives. The practice can be rather different. Most organisations claim to apply democratic principles, both internally in their staff management, and externally in their contribution to the good of society. Yet organisations, whether government, industry or social service, readily become internally referenced, walled off from the larger community, and distanced from their clients.

In many organisations, decisions are made solely with reference to the organisation's own short term goals, regardless of the social setting. This isolation arises from the management profession's claims to be both value and content free. That is, they hold a manager should hold no values or knowledge of their own, apart from their administrative skills. This neutrality, they say, enables a manager to work efficiently to any set of principles or knowledge base required by their employer. The approach has become known as outcomes-based management, or managerialism. From this perspective, managing a prison, an orphanage or a university is the same operation, run according to the same set of rules.

Rees and Rodley in their 2004 *The human costs of managerialism* report on the results of the failure of managerialism to distinguish between the interests of public and private good, and between those of the organisation and the larger society. The story in box 8.2 below *If the corporation were a person, would that person be a psychopath?* describes the outcome of taking this to extremes. On the other hand, there has always been principled collective management, and it is possible to predict that it is increasing, not decreasing. It is also possible to say that effective collective management is essentially democratic, in the original sense of serving all members of society equally. Box 8.1 holds examples of collective management in industry, research, administration, finance, education, and local and global governance.

Box 8.1

DEMOCRACY WITHIN ORGANISATIONS

- Alliancing
- Collaborative action research
- Deming's Total Quality Management
- Democratic governance
- Community banks
- Healthy cities
- Integrated management systems
- Learning circles
- LETS (local exchange trading systems)
- Learning organisations
- Sustainable communities
- World Commission on big dams

Management manuals have come to acknowledge the need for innovative organisations to develop alliances with all the interests involved in their affairs. The Whole-of-Community Action Web (figure 8.4 below) suggests six forms of connection are needed to hold communities together as they learn and change. Any organisation can use the same set of connections in becoming part of a forward-looking society. The connections are: bonding among members of the community; partnerships with close interests such as clients, consumers, funders, and the host community; alliances with other organisations; strategic planning involving the full set of decision-makers; and a set of ethical driving principles giving a focussed direction.

Management guidelines have been developed for each of these. Wenger's work on communities of practice, and Peter Senge's on developing a learning organisation, both report on ways of increasing the internal strengths of an organisation. Public/private and public/private/community partnerships are emerging as core business for most major cities. Alliancing has been the theme of a series of protocols being tested in project management. A global consortium, the World Commission on big dams, recommends an integrated organisational alliance as the basis for sound decision-making at the global scale. .

These tactical arrangements are directed towards greater effectiveness of the organisation itself. If the organisation is to contribute to a better society, this in turn will benefit the organisation in a feedback loop. For this, the organisation will need to embrace the larger ethical principles shared within that society. The most frequently invoked principle is that of a democratic society.

The switch from managerial to collective organisational management is in itself a step towards democracy. The father of the American democratic tradition, Thomas Paine, did not envision the organisations of democratic government as having uncontrolled power over society. He considered them the servant of society, acting as safeguards against potential conflicts of interest between individuals, their communities and the larger society. Paine's basis for the construction of a future, democratic society is hardly starry-eyed. In his classic treatise *Common sense*, he leaves the judgment of the successful application of democratic principles to the citizen and the community:

"Here then is the origin and rise of government; namely, a mode rendered necessary by the inability of moral virtue to govern the world.

> *Here too is the design and end of government, viz. freedom and security...................however prejudice may warp our wills, or interest darken our understanding, the simple voice of nature and of reason will say, it is right."*

Thomas Paine Commonsense

Democracy as a formal structure emerged in the late 18th century, almost simultaneously in Britain, the United States and Europe. The ideas emerged collectively. Edmund Burke, Thomas Paine and Francois Voltaire, the forerunners in each of these countries, acknowledged that they had drawn inspiration from each other's work and their interest was in the management of the whole of human society. All three expected populations to be small enough for individuals to balance the control and the protection of democratic freedoms through their own experience and commonsense.

None of these writers could have foreseen the human population explosion and the need for human beings to manage the entire planet. Nor could they have foreseen the arrival of the powerful global organisations without national affiliations or controls. Nor would they have expected to find giant international non-government organisations such as the World Health Organisation and the International Labour Organisation acting as agents for democracy. A coherent management framework can offer all these forms of organisation a working basis for implementing democratic principles.

The sociologist Ulrich suggests that there are four elements that underpin a society. Translating these for a democratic society, they would be shared principles, equitable power structures, a transparent knowledge culture and some means of social legitimacy. Democratic principles offer that social legitimacy. The present book proposes theory and practice towards a collective knowledge culture. The challenge remains to establish the other two elements for a collectively thinking society: the shared principles that make up a democracy, and the structures that move them into practice,

A reciprocal system leading to a collective ethic has been proposed by systemic thinker Richard Bawden. Such a system acts as an antidote to value-free managerialism. It also serves as a corrective to the competitive silos that have become the standard management structures of organisations. Bawden's holocentric organisation balances two sets of ethical dimensions often assumed to act in opposition to one another. The twin ethics of maximising personal liberty and controlling that liberty toward social justice are presented as mutually reinforcing rather than competing in figure 8.1. This is consistent with the classic essay *On Liberty* by John Stuart Mill. The twin ethics of organisational effectiveness and community well-being are set in a feedback loop leading to their collective advantage, the goal of most far-seeing managers. The two sets of ethics intersect at a point of harmony which sets the overall direction for a holocentric democratic organisation.

In figure 8.1, the interactions between the ethical principles form four quadrants of action. Each quadrant generates a need for an appropriate management structure. The American management expert James O'Toole, in *The executive's compass: business and the good society*, offers ideas for matching ethical directions to forms of organisational management. The structures are required to act as antidotes to what is currently the case, as well as to open the way to put the shared principles into action. Therefore the needs of a democratic, collective management system are for open collaboration rather than competition between departments; transparent alliances rather than hidden agendas; regulatory systems rather than poorly controlled exercise of power; and exchange-relationships as well as market forces. Providing the different administrative structures appropriate for each of the four quadrants represents significant structural change for most organisations.

In figure 8.1, the left hand top quadrant links individual liberty to community well-being. This relationship relies on voluntary collaborations, harnessing individual good will to enhance a sense of community. In the left-hand lower quadrant, where the task is to combine community well-being with social justice, formal alliances are required. These ensure open communication channels between disadvantaged groups and mainsteam organisations, so that the often voiceless groups can be heard. Consumer groups and social service lobby groups provide this function.

In the lower right hand quadrant, protecting social equity from too single-minded a pursuit of organisational effec-

Figure 8.1 The holocentric organisation: collaborative structures balancing the tensions between management principles (after O'Toole 1993 and Bawden 1995).

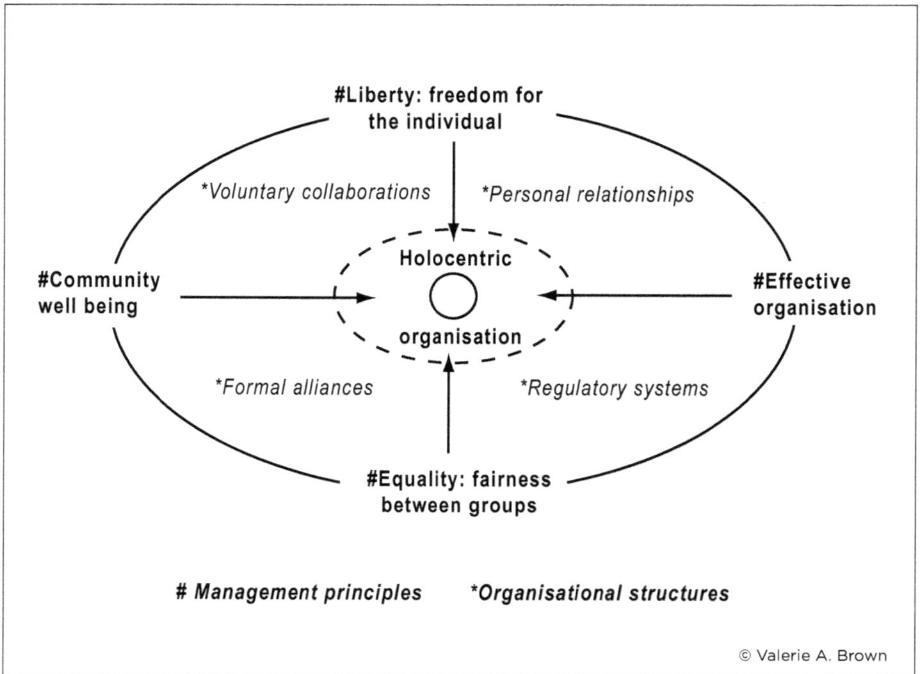

tiveness (and even more from mindless efficiency) means that regulations are required. These are increasingly taking the form of requiring due diligence and responsiblity for duty of care, The relationship between individual liberty and organisational effectiveness, the upper right hand quadrant, is a personal one. Maximising the quality of their interaction requires trust and respect between the actors. This trust is a feature of the democratic organisations such as sustainable communities and community banks, listed in box 8.1.

A key concern is for managers to keep in mind that this is a collective exercise. The objective is to achieve the harmony between the ethics, not to pursue one at a time. The same goes for the four management structures that will maintain the integrative system. Ensuring the presence of all four is more important then increasing the capacity of any one.

Holocentric, integrative management embraces two connecting democratic principles. One is Paine's proposition that the task of governance is to reconcile collective security with collective freedom (roots and wings again). The other is Voltaire's conviction that good governance is achieved through a marriage between an ethical direction and human reasoning, a combination that Paine and Voltaire both call 'common sense'. Within the competitive conditions of the market economy, holocentric organisational management is sometimes dismissed as inefficient and impractical. On the contrary, the examples described by the authors quoted above confirm that it can serve to improve the market system. On the basis that a holocentric organisation is a practical proposition capable of contributing to long-term social change, we proceed to explore the conditions which support it in practice.

What is:
conditions suppporting holocentric organisations

It is no simple matter to re-orient the power structures of a society to sharing a collective ethic. A clear and agreed framework for decision-making such as is supplied by the holocentric organisation is one essential. Another is to establish the enabling conditions that support the change. Otherwise, for the agents of change involved it would be like banging their heads against a brick wall. Not surprisingly, the Local Sustainability Project studies found an extremely high burn-out rate among those responsible for managing change. They also found that the key enabling factors were equitable power relationships, settings that supported the voice of civil society, integrative use of administrative structures, and committed leaders of the change.

Equitable power relationships

Until the insights of the Renaissance, power was fixed and personified by a single authority model - a god or a king. During the Renaissance, power became recognised as embedded in the individual and the institutions of the time. In *The Prince*, the Renaissance writer on power, Machiavelli, advises the prince how to use his power to manage the populace, not how to rule by divine right.

The Renaissance saw the rise of most of the institutional forms we take for granted today, many as formal guarantees for the first time. The charter of the city-state of Urbino guaranteed freedom of religion, one of the few in Europe that extended this freedom to its Jewish citizens. The Medici family of Florence opened the first public banking system, instead of simply following the tradition of hoarding their wealth. The guilds, forerunners of both unions and professional associations, operated social welfare organisations that supported the disadvantaged of the whole community, not merely their own members. Still far from democratic under autocratic rule, the Renaissance (rebirth) nevertheless produced these seeds of democracy.

John Ralston Saul is a major critic of links being made between value-free management and democracy. He concludes that any attempt by society to answer the question 'How shall we manage our affairs?' by logic and objective reasoning alone must be illegitimate. All management involves taking some ethical position. Hence in his 1995 book *Voltaire's bastards: the dictatorship of reason in the West*. Saul points out that almost all modern institutions claim descent from Voltaire's principles of democracy. These principles are based on harnessing the human capacity for rational thinking to the ethical principles of liberty and equality. Without these principles, no organisation, however efficient in its own terms, can justify claims to being democratic. In box 8.2, an account of the influence of value-free management over society supports Saul's criticism. The story is provided by *The Economist*, itself a premier organ of the market economy.

Organisational structures, whatever form they may take, exert their power through three avenues: the collective consent of the members, an implied promise of some improved future state, and control over the allocation of finite resources. This is the organisational context for the values of palace, temple and granary that hold a community together discussed in chapter 6. One of the most insightful writers on the combined effects of all three avenues of power is the French social philosopher Michel Foucault.

Foucault has documented the ways that the institutions of a society (prisons, hospitals, schools, families, and law courts) function as a coordinated interrelated system. He described how each such system can treat the same phenomenon quite differently. In different settings, homosexuality is treated as an illness, a crime, a personal choice, or a community concern. The first place where the spread of AIDS declined after its first diagnosis in the homosexual community was Sydney, where that community was given social responsibility for its control.

Box 8.2

IF THE CORPORATION WERE A PERSON, WOULD THAT PERSON BE A PSYCHOPATH?

An award-winning documentary film, *The Corporation* is based on a key 19th-century law that treats companies as persons under law. By bestowing on them the rights and protections that people enjoy, this legal innovation technically gives the company the freedom to do as it wishes. This allows a firm to act as singularly self-interested. Its purpose is solely to create wealth for its shareholders. It puts others at risk to satisfy its profit-maximising goal, harming employees and customers, and damaging the environment. It has no empathy, refuses to accept responsibility for its actions and feels no remorse. It relates to others only superficially, via make-believe versions of itself manufactured by public-relations consultants and marketing men. In short, if the law gives a firm the rights of a person, then the corporation is clinically a psychopath.

Lucy Hughes of Initiative Media, an advertising consultancy, is shown musing about the ethics of designing marketing strategies that exploit the tendency of children to nag parents to buy things, before comforting herself with the thought that she is merely performing her proper role in society. Mark Barry, a 'competitive intelligence professional', disguises himself as a headhunter to extract information for his corporate clients from rivals, while telling the camera that he would never behave so deceitfully in his private life. Human values and morality survive the onslaught of corporate pathology only via a carefully cultivated schizophrenia: the tobacco boss goes home, hugs his kids and feels a little less bad about spreading cancer. Company executives and foot soldiers alike will identify instantly with this analysis, because it is accurate.

Source: The Economist print edition, September, 2004

In *Discipline and punish* and *The clinic*, Foucault argues that an individual's social behaviour such as voicing dissent or a different child-rearing practice can be treated as wisdom, a crime or an illness, according to the controlling system of thought. The individual is then assigned to a university, a prison, or the health system, In each case, it is a communal agreement that assigns these ideas, so the people themselves accept their lot as inevitable. Law and government do not set the direction, they merely confirm it.

Foucault also points out that during the industrial era, access to knowledge was power. At that time, technical tools, new inventions, and professional skills were tightly held in guilds that controlled entry to their knowledge. In this information era with its open access to the information that builds knowledge and skill, he argues that access to the mode of construction of knowledge is power.

This perspective on a society as an interrelated dynamic system expands on the conclusion reached in chapter 2, with the story of changing the working day. Foucault's work illustrates the ways that changing any of the key elements of society impacts on all the others. This therefore offers some guidance on how all the elements must be considered in any design for organisational change. The studies reported here leave no doubt that bringing

Figure 8.2 Global and local socio-dramas and civil society (after Falk 1999)

	Globalisation from above		**Globalisation from below**
	unregulated flows:	Civil Society	locally regulated flows:
Flows of:			
Information	Information technologies		Networking local knowledge
Finance	Globalised corporations		Regional economies
Governance	Global institutions (World Bank, IMF, World Health and World Trade Organisations)		Local community networks, industries, organisations
People	Immigrants, refugees		Family unit, sense of place
Natural resources	Billion-year-old global air water, energy cycles		Local ecosystems, and landscapes
	National governments		Provincial and local governments

together the different power regimes of the knowledge cultures of Western society changes the overall dynamic of the society, as argued in chapter 3. A holocentric organisation may work to a distinctive collective ethic, but this still needs to be set in the context of open governance, integrative management structures, and champions of the change.

Open governance

The emergence of citizen-based civic management systems was a feature of the governance of the bounded city-states of the Renaissance. Governance refers to all the influences on the way a society is managed, of which formal government is only one. International relations analyst Richard Falk describes the context in which holocentric organisations must try to flourish in the 21st century. In his 2000 *Predatory Globalizations: A Critique*, he describes civil society as the negotiated territory between the two largely independent global flows of globalisation-from-above and globalisation-from-below. The interaction between the two as a factor in community change has already been mentioned in chapter 6. Here we are concerned with the ways holocentric power relationships can be negotiated across local and global scales.

The unregulated global flows of information, finance, governance, people and environmental resources are largely out of the control of nation states, although nations mediate them within their own societies. Falk calls these flows socio-dramas, and points out the many ways in which they have eroded the power of the nation state and increased the influence of civil society (figure 8.2). At the global scale, the organisations are largely invisible, and the result of their impact largely uncontrolled and therefore unpredictable. In contrast, local community organisations that are linked together world-wide, and so make up globalisation-from-below, have very visible communication strategies. There, the socio-dramas are locally regulated through local networks, as people simply get on with their lives. Locally elected community leaders and community activists who become engaged in issues are known first hand as individuals. State and local governments mediate the local flows in relation to the national interest (figure 8.2).

Civil society, which lies in the space between globalisation from above and below, is the place where the community negotiates between the two scales of government. Civil society is not bounded by those twin organisational structures, however, but can go direct to either global institutions or local communities. One example is the world-wide outcry when the global pharmaceutical industry refused to make available cheaply to developing countries the patented drugs on which huge profits had already been made, such as for AIDS and tuberculosis. Another is the rejection by some communities of the planting or the selling of genetically-manipulated crops produced by giant global chemical corporations. In this complex system of governance, the principles of holocentric organisation are core to the functioning of civil society.

A further example of a wicked problem requiring holocentric management principles is the response to the loss of the world's great forests. The results of the loss include displaced communities, changes in the global climatic patterns, local drought and floods, and loss of productive soils. Commercial tree plantations have been the solution to replacing forests favoured by most countries of the world. The story of the resulting impact is taken from Australian forester Jackie Schirmer's course on plantation forestry (box 8.3).

To fulfill the conditions for holocentric management, a collective solution to a wicked problem brings all individual needs into a single system of negotiations. In the case study of plantation forestry summarised in box 8.3, each knowledge culture selected its own avenue of action to resolve the issue. Plantation companies wanted to manage, governments to regulate, communities to set the agenda and specialists to set the benchmarks. The issues arise from tensions between company efficiency and community autonomy, and between individual freedoms and public controls. Both structures and principles are in a state of tension predicted by Bawden and O'Toole. As with all wicked problems, there is no final answer. The synergistic resolutions proposed in box 8.3 give reason for hope, if the knowledge cultures can learn to share their territories.

Integrative structures

A reason for optimism is that any organisation must have some integrative processes in place if they are able to function as a single organisation at all. Every type of organisation, therefore, holds the potential for holocentric decision-making. Figure 8.3 lists 14 of the many possible forms of organisational structure. It has been traditional in the analytic era to focus on the divisions, and define the problem as the conflicts between the resultant silos. But the ground rules for collective decision-making apply to any of the structures, although some are more sympathetic than others. In figure 8.3, the structural forms 6 to 14 offer more opportunities for synergy than do forms 1 to 6 with their firm boundaries.

Agents of change

In the field studies reported throughout this book, the single, committed individual is found to be the most likely catalyst for initiating the change from compartmentalised to holocentric management. In chapter 6 the characteristics of managers were found to match occupations and field of action. Thus each of the four structural supports for holocentric management will suit some personalities rather than others. According to the research findings of David Kolb and his colleagues, regulatory practices will suit convergent thinkers; setting up alliances will suit assimilators. Collaborations will best be managed by accomodators, and trust in personal relationships by divergent thinkers. This suggests the need for multi-skilling all members of

Box 8.3

A WICKED PROBLEM: TREE PLANTATIONS VERSUS NATIVE FORESTS (RESULTS OF d4-p4 WORKSHOP)

Industry: Tree plantation companies are usually internationally based, with none of their permanent employees living in the local towns. They are stock exchange listed, so must provide an economic return to their investors. To do this, they need to plant large areas of plantations. They consider that they are acting in the planet's best interests in providing trees as sinks for the world's carbon dioxide emissions; and in the district's best interests by increasing income. They argue that they should not have a compulsory code of practice for their plantations, since this does not apply to wheat or cattle.

Specialists have developed a code of practice based on principles of best practice sustainable management of forests. The scientists' advice is that instead of establishing large plantations with only one type of tree, the companies should manage native plantations which help restore biodiversity and environmental quality. However, they support the role monoculture plantations play in reducing logging pressure on native forests. They also assume that they will withdraw from management once their advice has been adopted.

For local communities, the impact is considerable. The economic viability of the district depends on developing timber production on marginal farming land. When farmers sell land for plantations, they often move out of the district. This leads to a loss of services – with every farming family that leaves, there are less children and schools are closed down; fewer customers and the bank branch leaves town. Neighbouring landholders find plantation companies 'bad neighbours'. They change drainage lines, so that water drains onto neighbouring properties and causes erosion. They don't look after boundary fences, and sheep and cattle escape into plantations. They haven't joined the local groups that shoot feral animals and safeguard local crops.

Local government initially welcomes plantations, because they will bring rates to the area. However, they later realise that the local government does not have the funds to upgrade the roads to support logging traffic. The plantation companies say that they pay rates, and so shouldn't have to pay any more for road costs. Local government does not welcome the prospect of having to administer one more contentious area without extra funds and possibly without community support.

A civil society solution negotiated among the interests above, was:

1. A compulsory code of practice to be supervised by Board containing key individuals and community, scientists, company and government representatives.
2. A code of practice to include action:
   International plantation companies to:
   - employ and train local farmers as a proportion of the plantation workforce;

Box 8.3 continued

- vary the monoculture by establishing large blocks of different species;
- practice good land management, maintain infrastructure, control feral weeds;

  and
- monitor actions and report against the regulatory code.

Farming community to:
- support farm forestry, providing traineeships, infrastructure;
- include pest control, fencing as part of farm forestry costs;
- increase species diversity for economic resilience; and
- monitor actions and report against regulatory code.

Local government to;
- lobby national government to allocate a budget to maintain roads;
- permit subdivision of farms for rural housing;
- set up collaborations between farmers, scientists, and plantations;
- subcontract local companies to log and remove trees; and
- produce State-of-the-region report on the combined actions.

Source of case study: Schirmer, J. 2000

a management team, as suggestec in chapter 5, or else putting together a multi-skilled team in the first place. Or a more collectively-oriented solution would be a combination of both, with members learning from each other and sharing their skills.

Advocates for collective practice within different forms of organisation work under many different labels. In community health, the usual term is a gatekeeper, applied to people who open doors to integration of social services. In multidisciplinary teams, one member may be called a tussock jumper, able to leap from discipline to discipline. Within an organisation, an innovator encouraging collective thinking may be called a pathfinder or ground-breaker. Each of these terms refer to establishing permeable borders among the knowledge cultures, and within their communities of practice.

Broader terms for agents of change refer to transforming the entire organisation or even the entire sector. One change agent in the Local Sustainability Project studies referred to 'organisational termites', people who gnaw away at the silo boundaries, reforming them as they go, until suddenly the whole organisation converts to integration, much as a termite-ridden building collapses into a single heap. A phrase coined by a bureaucrat who had managed significant change is 'the people in the cracks'. These are the integrators who loyally do the work of their segmented organisations without any open recognition of their larger goals. Eventually, times change

Figure 8.3 Integrative structures for holocentric organisations

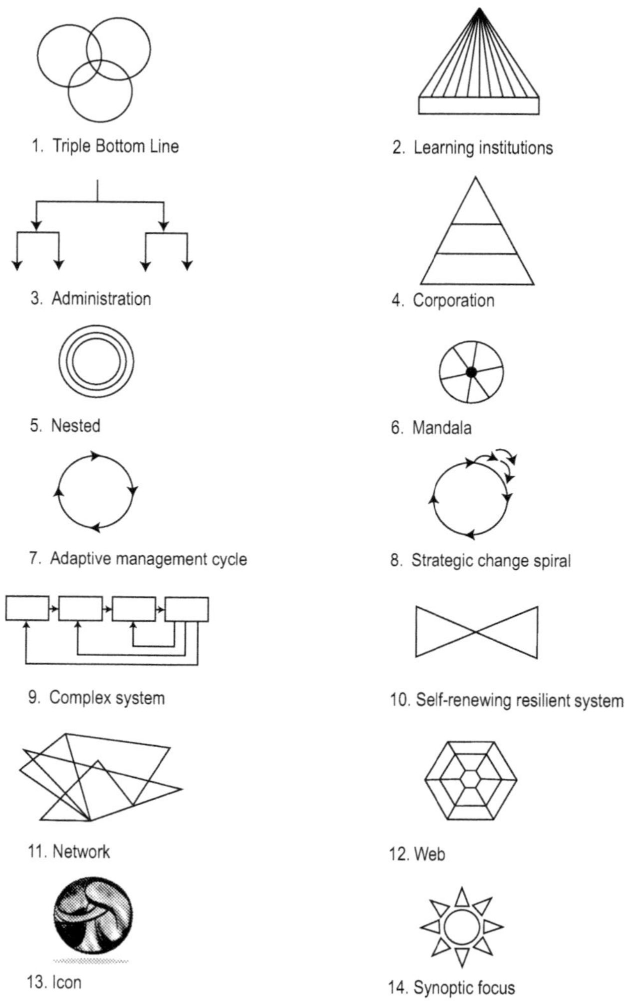

towards a more collective ethic. They then emerge from the cracks and dance, ready to work together for change to a broader agenda, such as sustainable development or social justice programs. Observers may be amazed at the apparent speed of the change, but it will have been in train long before.

The preceding discussion describes how holocentric organisations provide the structures to reconcile the ethical principles of equality and liberty, community and efficiency. The strategic planning that best drives the administrative structures in figure 8.1 towards being holocentric needs equitable power relationships, democratic

Figure 8.4 The whole-of-community change web

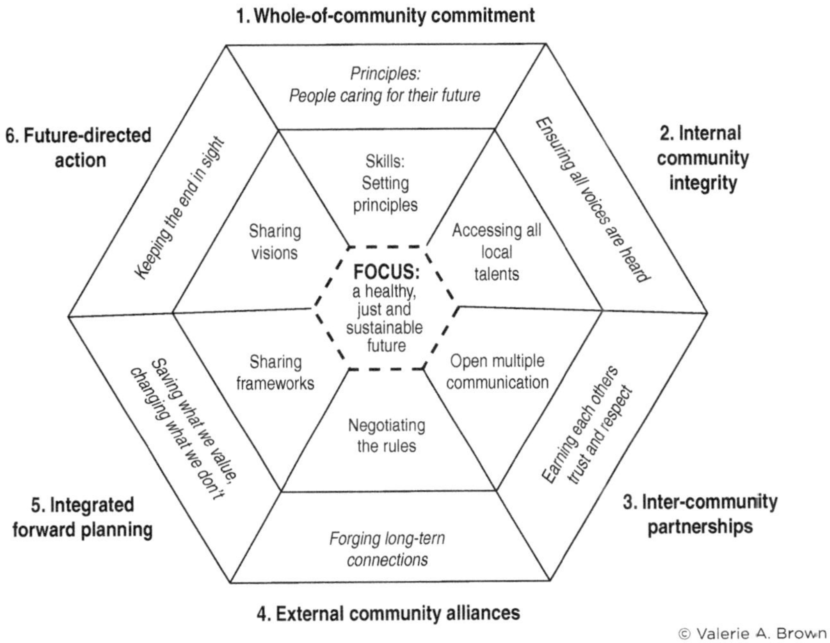

Source: Nicholson, R., Stephenson, P., Brown, V.A. and Mitchell, K. eds. 2002

forms of governance that include civil society, and committed agents of change. We proceed to examine how different forms of organisation can achieve these goals.

What could be:

whole-of-community action web

Environmental health offers a fruitful example of a system undergoing transformational change. As a profession, environmental health has already transformed itself twice over the past two centuries. The field emerged as a professional response to the health hazards that arose during the great 19th-century move from villages to cities. Its brief was to control the epidemics of cholera, diphtheria and tuberculosis under the crowded, unsanitary conditions. It was the branch of public health that advised engineers, health services, urban planners and politicians of the need to include hygiene in their professional practice. This type of environmental health has always been required to combine the knowledge cultures.

After the introduction of sewerage, enforced hygiene regulations and town planning had all but eliminated the epidemics, environmental health withdrew into a risk-avoidance profession.

During the 20th century it was charged with laboratory testing and policing roles. More recently, with the rise of the planetary risks to health described in chapter 4, environmental health has been under pressure to resume its socially sensitive role of pioneering a social transformation that it played in the 19th century .

The transformation from a maintenance role to a leading social change agent has not been without its stresses. National reviews of environmental health in Britain, Canada and Australia concluded there was need for major re-orientation. This challenge was met in the Australian context by a two-year consultation between environmental health practitioners from all the knowledge cultures: individual change agents, community activists, public health practitioners, government and non-government agencies and holistic coordinators. The resulting handbook produced by Rosemary Nicholson and others was titled *Common ground and commonsense.*

More than 30 national workshops were held using the same format as the one for social planners described below, with equal participation from all the knowledge cultures. The results of the workshops were synthesised in a synoptic workshop similar in design to the one that produced the text Sustainability and Health, described in chapter 7.

The outcome was a collective planning framework, a Whole-of-community Action Web that includes all the decision-making sectors. The Community Action Web links organisational structures into a single system, as recommended for the holocentric organisation. The sections, designed by those present at the workshop to a prototype developed by facilitator Peter Cuming, were:

- people caring for place: strengthening local ownership;
- communities in action: having our voices heard;
- communities as partners: earning community trust and respect;
- multiple alliances: forging long-term connections;
- place-based planning: saving what we value, changing what we don't; and
- future-directed action: linking environment and health.

Before publication as a handbook for guided social change, the whole-of-community action web was trialed in three locations. The governance structure, principles and actions from figure 8.4 were applied in a packed seaside suburb, a desert township, and an isolated indigenous community. Their wicked problems were 1. a pressing need for litter management in a national beach icon drawing 10 million visitors a year; 2. the local community's wish for a community park in an arid and isolated town; and 3. establishing an industry that could support an indigenous community. Five years later, each of these needs had been met, and a collaboration process was in place for other whole-of-community changes.

Overall, the web structure provided a firm and balanced structural platform in which collective action could take place.

What can be:
collective advocacy

In each of the ten field studies of the Local Sustainability Project, collective organisational change was initiated by a single committed individual, who then became the nucleus for an advocacy group. The agents of change came from every level of their organisation, from part-time project officer to chief executive officer. The range in age was between 22 to 65 years, and in occupation from scientist to administrator to community activist. The range of entry strategies was diverse, from overt advocacy to working quietly within the existing system, from mentoring others to taking open leadership. The starting point for change was thus anywhere in the organisation; what mattered was the commitment to the change.

There were some consistencies in all the sustainability projects, however. There were often partnerships between a young, enthusiastic advocate and a senior member of the same organisation supporting the change from a safe distance – an 'old dog, young dog' learning partnership. Another feature was a personal capacity for networking and negotiation, and yet another the skill of bridging boundaries. Often the agent of change was unaware of the strength of his or her own influence through these channels. Each of these strengths could become a weakness if, as occurred all too frequently, the change advocate became exhausted or discouraged, and left.

An important finding was that discouragement was often felt too early in the process. On several occasions the desired change occurred shortly after the proponent had left. In other cases, colleagues of the original change agent carried on the collective process in that organisation or another. Support for change agents that includes measures to reduce discouragement seems to be warranted.

The two case studies that follow describe the needs of advocates for organisational change from within government. One is from a single profession: 16 social planners from a range of government organisations seeking social change towards social justice. The other is a small group of advocates championing increased responsibility for sustainable development by the local government sector. In both cases, the advocated change was from traditional hierarchical, segmented management towards holocentric management,

Social planners

Taking a part in achieving organisational change has always been part of the duties of social, environmental and strategic planners, and public health practitioners. Each of these groups works to a strong and explicit value base for enhancing the public good. Each has at least some organisational licence to bring about transformational change. Each must, implicitly or explicitly, reconcile the priorities of individuals, communities, specialists and organisational strategists in order to establish change.

An open learning workshop was held for 16 social planners from a range of organisations in the same large capital city. Following the d4-p4 process recommended in previous chapters, the workshop was held in a room with four large round tables, each with a prominent label for the knowledge base of each of the decision-making sectors:

- Local knowledge: the community's lived experience of place.
- Specialist knowledge: the professions and academic disciplines.
- Strategic knowledge: ways of achieving organisational goals.
- Holistic knowledge: maintaining the essence, core theme or purpose.

Participants were invited to sit at the table labelled with the type of knowledge they normally used in their work. Only one person chose the community table, 4 went to the specialist and 4 to the strategic table, with 2 identifying themselves as using holistic knowledge. Three people placed themselves in between the specialist and strategic tables, and two between the holistic and the strategic, reflecting the bridging roles that social planners are expected to play.

Each of the four tables of knowledge types (local, specialised, strategic and holistic) were asked to identify the resources they use in their work, the power and authority which drove their work and the time-scale for successful completion. The responses, collated in figure 8.5, match those from the larger Local Sustainability Project studies in regional management and public health already described, 250 respondents in all. Together they give an overview of the conditions these participants considered necessary to move the strategic agenda from the separate decision-making interests to a more holocentric direction.

Advocacy group

An Integrated Sustainability Forum was made up of 25 local government practitioners who had already been recognised as national leaders in introducing sustainability programs into the local government sector. The sustainability champions played many different roles in their local government authorities, including member of the executive, strategic planner, director of environmental services, specialist advisor in community services, part-time project manager. Each participant could be described as both a gatekeeper and a boundary rider, with a commitment to opening the gates and breaching the boundaries.

The forum design included email discussion groups for three weeks prior to a two-day open learning workshop. A major finding from the workshop was that it was not so much any one tool or any one position in council that had won them success. Their leader-

Figure 8.5 Resourcing the knowledge cultures... what's on the agenda?*

Summary of basis for action:

1. Resources needed:

Community:	Networking, personal contacts
Specialists:	Funding for research programs
Strategists:	Negotiating strategies, legislative frameworks
Holists:	Collaborative team-building

2. Power and authority:

Community:	Local experiences, agencies they can trust
Specialists:	Discipline and research reputation
Strategists:	Legislation and political power
Holists:	Setting the direction

3. Timescale:

Community:	Action needed today, if not yesterday
Specialist:	Enough time for research to give a reliable answer
Strategists:	Enough time to negotiate the numbers, check the legislation
Holists:	Enough time to establish a coordinating agency

*n =16 social planners, 35 environmental planners, 200 in community workshops.

ship had arisen from their commitment to local sustainability, tools that matched their personal skills, and political opportunities.

The recommendations of the forum provided an agenda for managing social change as a core element of local government's conduct of its affairs:

- Changes in lifestyles, livelihoods and landscapes have led to a universal need for local knowledge, working from the inside out;
- Collective decision-making depends on a cooperative ethic throughout councils' activities not as an add-on extra;
- The full set of collective learning principles should be adopted as an essential overall learning directive for councils' activities, with responsibility accepted across the council organisation;
- Any integrative framework or tool can act as a successful entry point for collective progress towards local sustainability, if the principles are followed; and
- Change from within is needed to further the capacity for holocentric management in the local government sector.

Eight members of the original 25-member Integrated Sustainability Assessment Forum agreed to form the nucleus of a local government sector advocacy group. The forum's purpose was to act as a catalyst for sustainability across the entire national local government sector. In calling themselves the Baton Forum (the aim being to carry on the baton of

Box 8.4

LIVING WITH LEAD – A CASE STUDY FROM PORT PIRIE

This case study offers an example of the demands that social change places on industry, government and non-government management. Go to box 3.4 for directions for a workshop design that could generate a collective synergistic solution for Port Pirie.

In the early 1990s, the World Health Organisation and the National Health and Australian Medical Research Council dropped the recognised risk levels from 20 to 15 micrograms of Lead in a decilitre of blood for the population as a whole, and 10 micrograms for children. This was after 30 years of consumer and community lobbying. The consequences for a Lead-affected town have been as follows.

Port Pirie is a town with 5000 inhabitants and the biggest Lead smelter in the world. The town has enormous civic pride, like most traditional mining towns – with lots of community organisations, solid Victorian stone buildings and a large park. The smelter is the biggest employer in the town, followed by the health services and then the grain terminal (the surrounding country is wheat country). The town has a 12-member council, amalgamated from surrounding small rural shire councils. The council has gained its first 'green' member – previously the director of Port Pirie Community Services. The Mayor is the shipping manager for the Lead smelter.

The smelter has been in action for more than 100 years, but has recently become part of a global consortium. Prices are not so good for lead, and the mine is under administration. The company is considering moving the mine operations to the Mexico base, where there are fewer industrial safeguards and lower wage costs. This would spell the end of the town.

In the 'safe', upwind parts of Port Pirie, the children's levels are below the danger level. In the downwind sectors, levels are still found of 30 micrograms and over. Treatment is difficult and painful and it can take more than 30 years to reduce to the advised level. Only 50 per cent of parents have their children tested for Lead levels or visit the Environmental Health Community Centre, including those with known Lead blood levels over 10 micrograms.

A recent study found that mothers of Lead-affected children may deny the risk, even with quite severe symptoms in their children, because Lead is an outcome of their husband's work, and they see no way of escaping or changing things. One desperately concerned mother with a child recently diagnosed with a level of 20 micrograms said: "My husband works in the mine, his father is the union organiser for the mine. I can't even mention it at home. Since our house has been notified as in the risk area for the new lead levels, it's worth nothing, so we can't move."

What are the issues for the strategic management of Port Pirie? Any expensive solution, such as compensation for mine families so they could move, or better technology that would eliminate the lead in the air, could tip the decision on the mine closing, and put two-thirds of the population out of work. Any political lobbying by the community health centre is going unheard. Those most at risk wish to remain silent, and there is only one social interest member of the council. There the situation has rested ever since the policy on permissible Lead level changed five years before.

Postscript: Ten years later

A new CEO has been appointed to the mine, one whose personal and organisational brief includes being connected to workforce and community. At the same time, a consumer action group has been ormed among mothers of Lead-affected children. Synoptic workshops linking these two ends of the power system have changed the perspective of the local Council, and alerted provincial and national governments. A medical researcher has been commissioned to construct a detailed social map to publicise Lead levels. The new international links of the mine have given it access to clean air controls in smelter technology. Local, state and national agencies are combining funds to re-position half the town. Once the pieces were connected, the community proved to have the internal capacity to manage its problems. Port Pirie seems back on track to being a good place to live.

Source: Brown VA, Field study, 2003

working for sustainability), members described themselves as a group of local government sustainability practitioners. They were explicitly committed to working both from the inside out, as local practitioners, and from the outside in as a national lobby group, at the same time (figure 8.6). They explicitly rejected the more usual top-down, bottom-up model of the power system transfer of power.

The Baton Forum's first step was to commission a review of successful advocacy organisations. Recommendations that emerged from the study were to:

- combine a broad range of interests and expertise;
- recruit some high-calibre, high-profile members;
- start by identifying aims, outcomes and resources;
- build a widespread network;
- try for an early success, and 'making a name';
- draw on specific expertise whenever needed;
- establish good relationships with relevant media; and
- reach a good political understanding, while politically unaligned

At the time of writing, the Baton Forum had prepared an ambitious forward program that included holding a national conference, developing an independent public voice, and sponsoring a mentoring program within the sector. It is an example of what can be in terms of organisational transformation from the inside.

In a holocentric organisation, the task

Figure 8.6 Relationships underpinning A. hierarchical and B. holocentric management (adapted from Brown VA in Hargroves and Smith 2004).

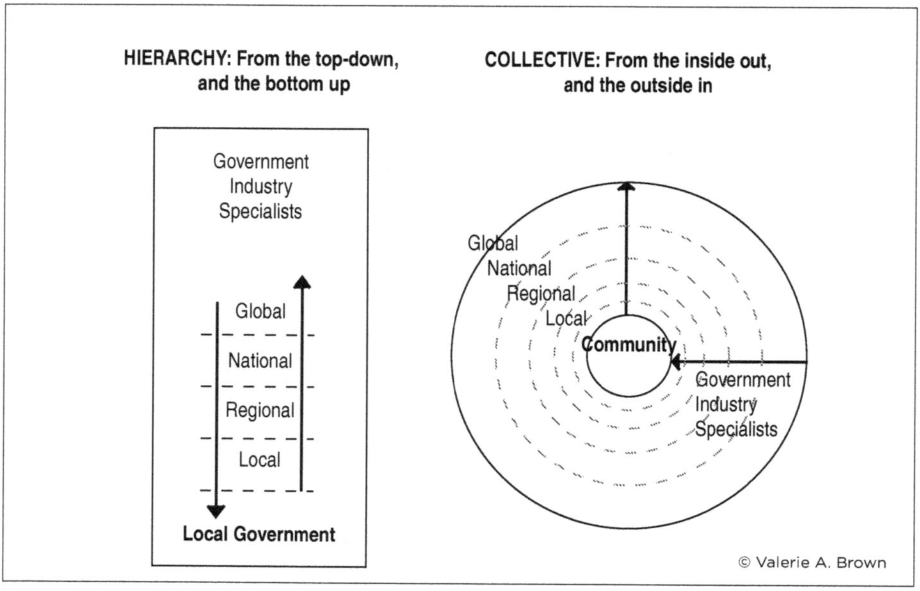

of management remains focused on synthesis as the main goal (*what should be*). The role of the organisation's strategic knowledge is to harness (*what is*). The power of such a synthesis, when working effectively, is apparent from the examples of the Renaissance - whether it be Machiavelli's prince or that acclaimed model of civic organisation, the city of Urbino. And, as the Medici family proved, breaking with traditional institutional norms can lead to significant private and social power (what could be). These examples act as ongoing lessons for today's community leaders, who are trying to establish what can be.

This chapter finishes with another case study from the wicked problem of Lead in air. This time it highlights the issues involved when the only solution to a social and environmental problem is through a collective decision to act. The selected issue is a community with Lead-affected children in a town where all forms of organisation depend on Lead production (box 8.4).

In the next chapter, there are descriptions of the role of the fifth knowledge culture, holistic knowledge, in collective thinking and action. Chapter 9 explores what it means to the other knowledge cultures for them to incorporate a holistic construction of knowledge. It also reviews a strategic tool employed by architects, professionals and computer software designers to acieve a holistic focus to their work: designing a pattern language.

9 A holistic focus

"To see a world in a grain of sand,

And a heaven in a wild flower,

Hold infinity in the palm of your hand

And eternity in an hour."

William Blake,
Ages of Innocence

The city of Paris, Earth from space and a smiling Nelson Mandela are all symbols for a far more complex understanding than their actual physical image conveys. The beautiful French city is an icon for a civilised life. The green ball in the sky conveys the fragility of humanity's only home. The smile of the South African leader stands for personal dignity and integrity under adversity. Each symbol is both a part and the whole of an idea at the same time. Leonardo da Vinci's writings and paintings reveal an obsession with conveying the actual and the universal in the one image. It is this human capacity to recognise the essence of an idea through a powerful symbol that gives us our holistic knowledge that allows ideas to be shared across time, and across cultures.

What should be:
discovering the core of things

Everyone is a holist, at least to some degree. To think holistically is the most essentially human, and possibly the least understood, of all human abilities. Whatever the chaos of our individual lives, or the complexity of the world in which we live, we are programmed to make sense of it as best we can. Intuition, imagination, trust in our own judgment and suspending disbelief enable us to create a fresh whole in our minds from familiar ideas. Innate common sense ensures that we 'pull ourselves together' under even the most demanding conditions.

The gestalt exercise in chapter 5 demonstrated that making sense out of even quite unrelated information is a human biological given. But there is much more to holistic knowledge than that. The creative leap that releases a fresh holistic insight belongs to the individual. For that leap to make a unique contribution to public knowledge is a more complex matter. The examples in box 9.1 remind us of just how powerful that leap can be.

Box 9.1

CHANGING THE WAY THE WORLD THINKS ... EXAMPLES OF HOLISTIC KNOWLEDGE

- Leonardo's paintings
- Picasso's *Guernica*
- Armstrong's broadcast from the moon
- Watson, Crick and Franklin's double helix
- McDonald's golden arches
- Geldof's Live Aid
- Newton's falling apple
- Archimedes' bath
- Hiroshima's atomic bomb
- Watt's steam engine

New York art critic Robert Hughes identifies *The shock of the new* as the contribution an artist makes to a whole society's social learning. Leonardo da Vinci's painting *The annunciation* (figure 9.1) carried just such a message of change to the community for which it was painted. Mary and the angel who announces to Mary that she will give birth to Jesus are painted as real people, not formal figures as previous painters presented them. Mary is not a simple countrywoman,

Figure 9.1 Leonardo da Vinci, *The Annunciation*, **Florence, Uffizi Gallery.**

as in the biblical account of the birth in the stable. She is richly dressed, the daughter of a great house, leaning over a book. The painting carries vivid detail of the Tuscan landscape, one of the earliest paintings to include such realism. Integrating it all is our ability to enter into Leonardo's holistic vision of a classic moment, the human shock of recognition of something unprecedented.

Before we can further understand the contribution of holistic knowledge to collective knowledge, we need to clarify the competing uses of *holistic*. Its inventor, the South African Field Marshall Johan Smuts, a polymath, soldier, scientist, philosopher, businessman and statesman, used holistic to refer to the essence of a whole, to the core of a dynamic system that goes beyond the sum of its parts. To quote Smuts, writing in 1931: "*Instead of the animistic, or the mechanistic, or the mathematical universe, we see the genetic, organic, holistic universe*" (J.C. Smuts, London Times, 7/7, 2.9.31).

Within the specialist tradition, a common interpretation of 'holism' is 'all the parts of the whole". The whole in this sense, whether the subject is a city, a planet or a person, is represented in standard analytic thinking as a set of interacting parts in a cycle or a dynamic system. Holistic thinking then becomes limited to a zero-sum game in which the whole is equal to the sum of the parts. Interpreting holism as merely the sum of the parts serves to preserve the dominance of the analytic tradition, and to exclude the potential offered by synthesis.

Holism as in searching for the essence of a system, a place or a person, is the sense of Smuts' definition when he originally coined the word. The structuralist Roland Barthes in *Natural symbols* distinguishes between the pointed to (the physical object) and the standing for (the symbol with a larger meaning). In interpreting community in chapter 6, temple, palace and granary stood as the pointed-to symbols standing for spiritual, temporal and economic power.

Cartoonists and painters have a particular capacity to capture such essences visually; writers and poets capture them verbally. We start our search for the role of holistic knowing in the collective construction of knowledge by looking for criteria for what it means to think holistically. We then explore the tacit use of holistic thinking in each of the knowledge cultures. We go on to consider the work of Christopher Alexander, an American mathematician, architect and designer who has spent a lifetime exploring the capacity for holistic thinking in designs for living he called pattern languages. Finally, we examine a study of holistic thinking in a range of professions, as possible role models for everyday practice.

Holistic thinking did not begin in 1931 with the coining of the word. It has been a mark of human existence in every society, culture and era, as in the examples in box 9.1. For English scientist Newton, watching an apple fall crystallised his understanding that gravity is a force common to Earth and the solar system. Spanish artist Picasso's Guernica brings together the elements of the first aerial bombing of a defenceless community so vividly that it continues to convey, more than any set of statistics, the horrors of all forms of war. The Greek philosopher Archimedes understood the relationship between mass and weight, as he watched his body replacing the water in his bath. His cry of Eureka ('I have found it') still echoes through the ages as the symbol for holistic thought.

Arthur Koestler, a holistic writer and thinker, examined the creative leap that extends the boundaries of knowledge in his 1978 *The act of creation*. He identified three holistic modes of recognising such a transformation. 'Aaaaah' is the gasp of breath from the shock of a fresh and unexpected insight. 'Ahah' is the shout of triumph at the moment of discovery of an unexpected idea or a solution. 'Hahaha' is the gurgle of laughter that comes with the twist of meaning that makes up a joke.

Some truths are uniquely affirmable in the symbolism of art. Paintings, poems and music can express holistic ideas inexpressible through any other medium. Art critic Donald Morgan vividly evokes the gap between an overwhelming holistic experience from listening wholeheartedly to great music, and a logical evaluation of the same sensation. In his article 'Must art tell the truth?', in the *Journal of aesthetics and art criticism*, he writes:

"Remember that breathless final moment when you have moved intensively with heart and mind through a quartet of Brahms or Bartok. You have hoped, expected, feared, been lifted, lowered, fulfilled, and disappointed, and now, inevitably the voices together sing one rich, climactic chord. You as a person vibrate, suspended with that one vibrating sound ... Then someone asks, what did you learn from that?"

The listener crashes down to earth from the holistic high of their intense experience.

Box 9.2 contains samples from writers working within each of the knowledge cultures. The poet gives himself greater

Box 9.2

CAPTURING THE ESSENCE: HOLISTIC EXPRESSION OF THE HUMAN-ENVIRONMENT RELATIONSHIP

The individual

I breathe in the only happiness I can attain – an attentive and friendly awareness. Young women looking after a children's holiday camp, the trumpet of the ice-cream sellers, the displayers of fruit, red melons with black pips, the black stickiness of grapes – are all props for the person who can no longer bear to be alone.

Albert Camus, 1937, *Betwixt and between*, p162.

The community

The village to which our family had come was a scattering of some 20 to 30 houses down the southeast slope of a valley. Living there was like living in a bean-pod; one could see nothing but the bed one lay in. Our horizon of woods was the limit of our world.

Laurie Lee, *Cider with Rosie*, p245.

The ecology

While a deer pulled down by wolves can be replaced in two or three years, a forest pulled down by too many deer may fail of replacement in as many decades. The cowman who cleans his range of wolves does not realise that he is taking over the wolf's job of trimming the herd to fit the range. He has not learned to think like a mountain.

Aldo Leopold, 1989, *Thinking Like a Mountain in A sand country almanac*, p132.

The organisation of society

I made my way under a downpour past great silent factories, Bleston, under great smokeless chimneys, past great closed iron gates and grilles, as far as the muddy walks of Easter Park ... I seemed to hear the murmur of all your sleeping inhabitants, Bleston, who long for your death from the very depth of their bones.

Michel Butor, 1960, *Passing time*, p20.

The holist

A child said 'What is grass?' fetching it to me with full hands ... it seems to me the beautiful uncut hair of graves,
Tenderly will I use you, curling grass,
It may be you transpire from the breasts of young men,
It may be that if I had known them I would have loved them,
It may be you are from old people, or from offspring taken too soon out of their mothers' laps,
And here you are the mothers' laps.

Walt Whitman, *Song of myself*, pp27-28

freedom to express his meaning with the use of symbols, but each writer employs holistic imagery. Each in turn contributes the essence of the human-environment relationship to their particular knowledge culture. Each introduces the holistic understanding of that relationship into the mind of the reader. The reader's recognition of the meaning behind their symbols reaffirms the statement at the beginning of this chapter: we are all holists.

In the 1969 compendium *Aesthetics,* philosopher John Hospers summarises the criteria for judging contributions to knowledge from the creative arts. Each piece of writing in box 9.2 meets the criteria for a contribution to holistic understanding, as set by the arts world:

- not as true or false, but having integrity and quality;
- not as insipid or pale, but vivid and colourful;
- not as weak or timid, but forceful and strong; and
- not as mild or neutral, but tender, ironic, tragic, graceful, delicate, and richly comic.

These criteria can also be applied to the work of Christopher Alexander, the American architect and influential thinker of the post-industrial era, in using patterns as designs for living. He has given the evaluative label 'having life' to any well-designed human product or structure. He sees the world as a series of centres, each the core of a pattern. Each pattern holds a rule for making centres. Strong centres are holistic and have life, and so contribute to a good society. Weak centres result in patterns without life, which he regards as socially destructive. He writes in *The nature of order: book 1: The phenomenon of life: "Whether it is an apple-orchard, a family, a turkish carpet, or a fervent kiss, in every case the life which that thing has arises from its wholeness."*

Alexander began his work with patterns with an analysis of the geometry of Turkish carpets, and went on to design whole cities. The first thing to notice in his list of characteristics of wholeness or life is that they are not meant to be separated, but considered equally important defining qualities. Life is derived from the mutual presence and the synergistic support for each other of all the qualities in box 9.3. The second point is that each criterion for wholeness encourages maximum freedom, not a limitation.

A third point is that the characteristics of a 'good' pattern apply equally to social and environmental change processes. Indeed, they almost always involve both. Alexander notes that the characteristics of a strong centre are easy to identify in natural systems, but harder to achieve in social systems.

The fourth and final point is that the list of criteria for strong centres can equally well be applied to the patterns of nested and networked knowledges explored in chapter 3. Where they produce strong centres, the resulting knowledge will 'have life', as in the examples in chapter 3 of collective thinking that uses nested knowledges. Alexander has applied his propositions (they are too practical to be called theories) to transforming urban planning, public spaces and public buildings.

What is:
holism in the arts, physics and ecology

Holism and the arts

Only in the creative arts is holistic thinking accepted as a dominant and desirable mode of constructing knowledge. In our analytic Western society, to be imaginative in any of the fields of specialisation is likely to be considered impractical and unreliable. In an organisation, it can be labelled as disloyal and losing the plot. In a community, it is a sign of difference and is controlled in the many ways each community deals with difference.

Artists have been called 'the early warning systems of mankind' for their capacity to use the power of holistic thinking to convey a shift in thinking ahead of the general citizen or the political system. Yet artists are subject to the same social control mechanisms that dismiss creative thinking in other fields. In 1914, the work of Michael Duchamp, a French painter, poet, technician and chess player, gave the viewing public the capacity to critically review the 'modern' products of the industrial world. This was long before social scientists were calling this type of critique 'post-modern'. In a series of works called 'ready-mades', he startled a public used to art as a record of the past and present. He introduced the realisation that art could also re-order the present and predict the future.

'Ready-mades' were practical objects of everyday use, such as a bottle holder, a snow-shovel and a urinal. Duchamp signed them as if they were an artwork, and gave them titles far from their everyday use. The urinal was labelled 'Fountain' (figure 9.2). The art world complained that Duchamp had signalled the death of art. Rather, these

Box 9.3

CRITERIA FOR STRONG CENTRES THAT GIVE LIFE

Criteria	Examples of patterns
1. Differing levels of scale:	space, time, structures, people, fractals.
2. Strong centres:	centres of action, nodes, nuclei.
3. Fluid boundaries:	zones of interaction, webs, river banks, edges.
4. Alternating repetition:	lattice, orbits, forests, ripples.
5. Positive space:	the space within the lines, bubbles, glades.
6. Good shape:	centres within centres, leaves, shells, hands.
7. Local symmetries:	volcanoes, human body, a fingernail, stars.
8. Deep interlock:	interacting surfaces, overlapping systems.
9. Contrast:	male-female, up-down, positive-negative.
10. Gradients:	regular variations of colour, pattern, strength.
11. Roughness:	irregularity, unevenness, dappled.
12. Echoes:	repetitions of key characteristics.
13. The void:	gaps and spaces that define systems.
14. Simplicity and inner calm:	simplest form consistent with purpose.
15. Non-separateness:	recognition that every system is connected to each and every other system at some level.

Source: Christopher Alexander, 2002, *The nature of order. Book 1. The phenomenon of Life.*

works led people to see their mechanical, industrially designed everyday objects as if for the first time, and to reflect on the meaning of modern and the 'modern' society that had produced them.

Figure 9.2 Michel Duchamp's 1917 *The Fountain* **Philadelphia Museum of Art**

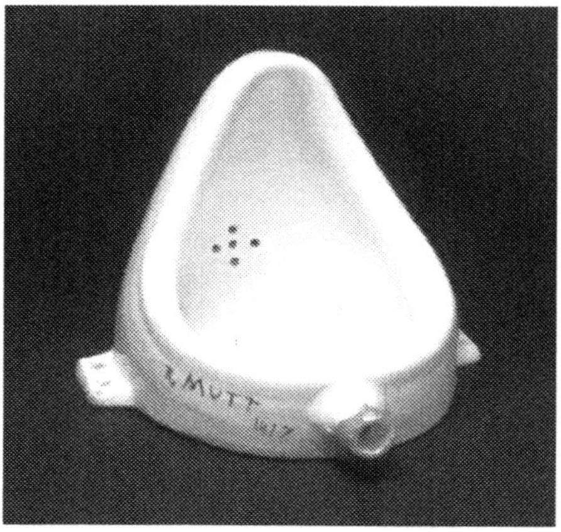

© Photo SCALA

The next step is to consider whether specialists use holistic thinking to focus their own disciplines so that they point the way forward for society as a whole. We find that they do, in much the same way that art does. Such a holistic focus is employed by scientists expanding chaos theory, biologists predicting the way for sustainability, and computer program designers extending relational communication systems.

Holism and physics

Scientists John Briggs and David Peat give a holistic interpretation of the social implications of chaos theory in their 1999 *Seven Life Lessons of Chaos*. They observe that the linear cause-and-effect, two-dimensional Newtonian view of the world has been found to be flawed so frequently that it has generated a crisis of confidence in science and expert knowledge as a whole. Chaos theory, originally used to predict air flows in thunderstorms, is now being applied to a fresh interpretation of the whole, for everything from medicine to social reform.

Chaos is now understood to be, not a formless unpredictable state, but essentially holistic. An underlying connectedness has been discovered in even the most apparently random events. Chaos theory directs researchers to the study of whole fields, acknowledging the impossibility of separating systems from each other within the same field. The fields themselves are also assumed to be connected. Briggs and Peat's *Seven Dimensions of Chaos Theory* links the science of chaos to the structure of society. The parallels these authors draw between social and scientific constructions of reality are described in box 9.4. Blake's poem at the head of this chapter foretells the concept of fractals in chaos theory. Alexander's 25 characteristics of 'life' has informed new directions in architecture and in computer program design. The idea of decision-making on complex issues as a set of collages in a learning spiral is a holistic interpretation of the construction of knowledge as a whole.

Holism and ecology

The ecologist/anthropologist/philosopher Gregory Bateson writes in *Steps to the ecology of mind* that social reality is constructed from within the current level of understanding of the biophys-

Box 9.4

NOT SO RANDOM ... SEVEN DIMENSIONS OF CHAOS THEORY

1. Creativity

Chaotic systems are controlled by a combination of negative feedback loops that reduce uncertainty, and positive feedback loops magnifying uncertainty. Together they eventually move to a self-organising system. An air conditioner is the simplest form, Earth's atmosphere one of the most complex. Knowledge of the ambiguities in such a system will be more accurate in predicting an emerging pattern than the most detailed description of an existing state.

2. Connectivity

In a chaotic or uncertain system, small changes can bring unexpectedly large results, labeled by the authors as 'butterfly power'. The fall of the Berlin Wall, the public discovery of the Watergate conspiracy and the chemical synthesis of the contraceptive pill are all one-off events that have changed the social and environmental patterns of the planet forever.

3. Self-organisation

This is the productive principle of both Earth's self-sustaining systems and cultural long-term survival. The physicist turned philosopher, Fritjof Capra, argues in his 1998 *Web of Life* that self-organisation is the primary characteristic of 'life' on earth. James Lovelock set up the official test of the US space program for 'life' in Mars as any evidence of self-organisation.

4. Intermittency

This is the scientific term for simplicity and complexity alternating so fast that they are indistinguishable, and virtually co-exist at the same time. This appears at first sight to be a paradox, but on further examination turns out to be a permanent condition of matter and of social events. Lights as both particles and waves, and trees as individuals and as forming a forest are examples of the general rule.

5. Living within fractal space

In chaos theory, this means living simultaneously with microcosm and macrocosm: the whole in the part and the part in the whole. Blake describes this duality as "the world in a grain of sand and eternity in an hour".

6. Living within fractal time

This means abandoning dependence on the linear backwards and forwards construction of time, established by centuries of human reliance on calendars and clocks. Time is multi-dimensional, and the continuous small changes in rhythm are indicative of living systems.

7. Holism recovered

Reality as a series of parts making up a predictable whole has been challenged in each generation, by artists, musicians, writers and scientists. In the last half of the last century the challenges have amounted to a review of the nature of reality. This is now increasingly regarded as a holarchy, a pattern of interconnected and inter-related wholes, each influenced by the others, and each greater than the sum of their parts.

Source: Adapted from Briggs J. and Peat, D., Seven Life Lessons of Chaos.

ical world. We noted above how chaos theory has changed the way we expect to interact with the world. Tim Flannery in *The future eaters* argues that it is biologically inevitable that each human generation will use its knowledge and skills to exceed the limits of the available biophysical resources, leading to hunger and war. The next generation will then need to use its creativity to reinstate a stable sustainable world. The capacity for holistic thinking has always provided the choice to escape this fate. The Neanderthals of Europe and the current barren state of Easter Island are examples of cultures that failed to make a re-adjustment. The productivity of Western Europe and the creativity of the Indonesian island of Bali are examples of cultures that have succeeded.

Using the tool of holistic thinking, scientists have interpreted the future-eating species' struggles with its partner 'nature' in metaphors that allow non-specialists to understand them. Martin Wark has documented the changes in a time series of interpretations of 'nature'. First nature is a planet untouched by humans, second nature when small patches are changed by human tools. Third nature sees humans with high technology changing the whole surface of the planet. We are now in fourth nature, living with the emerging awareness that we are now the custodians of the planet for future generations (figure 9.3).

Jared Diamond, an evolutionary biologist turned socio-biologist, identifies the human species as one of three evolutionary branches of the chimpanzee line. In *The rise and fall of the third chimpanzee* Bill Rees, co-inventor of the holistic idea of the ecological footprint of human impact on the earth, coined the phrase 'patch disturbers'. Arthur Koestler identifies the understanding of the role of humans during the industrial era as 'the ghost in the machine'. Tony McMichael, a leading researcher on humans and global environmental change, took as the theme for his seminal book *Human frontiers, environments and disease* a more hopeful direction for humans as stewards of the future.

In this era of human-generated social and environmental change, experienced scientists are abandoning sole reliance on technical definitions and moving to convey the significance of their work with the aid of a holistic symbolic language, as in figure 9.3.

What could be:
a pattern language for civic communication

The aim of Christopher Alexander's life work has been to generate synergistic patterns of interaction between humans and all aspects of their environment. His scope includes cities, houses, parks, natural landscapes – anywhere that humans construct their lives. In four seminal books Alexander puts together several hundred such patterns, all interconnected. He writes that together they form a language. The language can be a medium for prose or for poetry. In prose, each word is expected to have one meaning, and each sentence one idea. In a poem and a pattern, each word carries several

Figure 9.3 Holistic interpretations of socio-biophysical nature of 'Nature'

Stages	Years before present		People and Place
1st Nature First humans	1 billion	2 million	Global bio-physical systems *The third chimpanzee* *(Diamond 1991)*
2nd Nature Humans with tools	2 million	300	Local bio-social systems *Patch disturbers* *(Rees 1991)*
3rd Nature Human tools shape the planet	300	Present	Local systems become global *Ghosts in the machine* *(Koestler 1965)*
4th Nature Rethinking human/environment relationships	Present	?	Whole-of-planet management system *Earth's stewards* *(McMichael 2001)*

meanings. Each sentence or set of patterns in a pattern language also holds an enormous density of interlocking meanings, which illuminate the whole. In this way, patterns are able to express the equivalence of the part in the whole and the whole in the part. Patterns, in Alexander's thinking, offer an intellectual collage of a fully connected world.

As an example, take the connections Alexander makes between water, as in a bathroom, and water, as in the landscape. He approaches the first as the part of the house where you can bathe yourself slowly, with pleasure, perhaps in company, rest your limbs and relax. The other is a place in the neighbourhood, where there is water in which to gaze and swim and for children to splash about. Each use of water nourishes a part of ourselves in which water is one of the elements of our evolutionary memory.

Another example of connectedness is the pattern language created by the needs of old people in Figure 9.4. Here Alexander links physical, mental, social and creative needs directly to patterns that address that need. Any design that links the private with the communal, and the socially constructed with the biophysical given, makes a powerful contribution to 'life'.

Douglas Schuler, a computer professional and academic from Seattle, is seeking to take Alexander's pattern language into a language of social change. He describes his project, *Liberating Voices: a pattern language for civic communication:*

"The aim is to identify patterns that can help us marshal creative, intellectual, and moral principles into synergistic information and communication systems designed for open learning ...".

As Schuler and his colleagues define their terms, they describe a pattern as 'a careful description of a perennial solution to a recurring problem, considered in its social context'. A pattern language communicates those solutions through its own grammar of label, issue, context, and solution. The term 'living communication' implies connectedness to the world through a language intrinsic to everything we do. The terms can be examined in more detail on the website http://cpsr.org/program/sphere/patterns/

While the aims of the pattern language project may be the utopian dream of a common human language incorporating all ways of knowing, the execution is practical. Designers and users come from almost every field of knowledge - science, management, art, politics, law and mathematics. There is a parallel between the theme, issue, context and solution of a pattern, and the sequence of principles, parameters, potential and practice of the decisions-into-action learning spiral. The network of contributors to pattern language development is spread over 16 countries, and includes practitioners from all knowledge cultures. The goal is to provide a collective vehicle for agents of change to work together in changing the society in which they live.

Each pattern can be compared to a sentence in standard language forms. A set of patterns forms a story in the pattern language, as in Alexander's pattern of patterns on aging in figure 9.4. Schuler gives a pattern for civic intelligence (figure 9.5), which offers a holistic focus for civic society communication between global and local institutions, identified by Falk in chapter 8.

Figure 9.4 Alexandrian pattern language: designs with life (after Alexander, 1985)

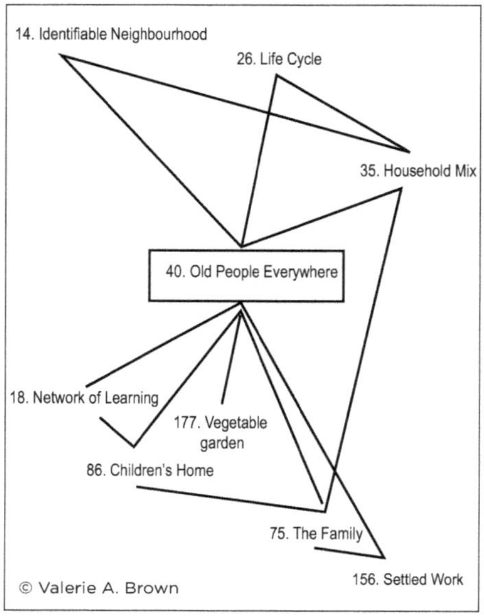

The use of pattern language in guiding social change is still in its infancy. However, when successful, it promises to deliver the conditions required by Christopher Alexander for patterns 'having life' (figure 9.4). A pattern incorporates the ideas of wicked problems, discourse and collective learning that we identified as essential elements of guided social change in chapter 4.

What can be:
pattern language as holistic practice

Alexander's ideas have been put into practice in a doctoral research project examining guided social change. In 2005, town planner Gregory Paine completed a doctoral PhD thesis on *Wholes, patterns and sustainable development: closing the gap between concern and action*. Paine was a member

Figure 9.5 Excerpt from a pattern for civic intelligence (Schuler 2006)

Problem

Humankind and the natural environment face profound challenges that will require an immense amount of attention if the challenges are to be reconciled appropriately. Society often develops intelligent collective responses to collective problems, but finds it difficult to put them into effect.

Context

Civic intelligence, like the various types of intelligences identified by Howard Gardner in *Frames of mind* can be used to explore a flexible and powerful competence that goes beyond the traditional notion of intelligence. But civic intelligence goes beyond the individual and focuses on the collective and distributed nature of intelligence.

Solution

An organization or other group that exhibits civic intelligence can be characterized using a descriptive model that contains six main components:

- orientation: the purpose, principles and perspectives that help deployment of civic intelligence;
- organisation: the structures, methods and roles by which people engage in civic intelligence;
- engagement: the ways in which civic intelligence is an active force for thought, action and social change;
- intelligence: the ways that civic intelligence is manifested through learning, knowledge formulation and sharing, interpretation, planning;
- products and project: the ways that civic intelligence organizations focus their efforts; and
- resources: types of support that people and institutions engaged in civic intelligence work need and use.

of the synoptic seminar group of the Local Sustainability Project described in chapter 5.

Part of Paine's inquiry involved interviews with practitioners from ten different professional fields (listed in figure 9.6). The participants were selected on the grounds that they considered themselves to be, and were also described by others, as 'holistic practitioners'. The aim was to discover whether there was a connection between professional expertise, holistic thinking and the use of a pattern language. The inquiry method combined two exploratory techniques designed to uncover key ideas, and to generate alternatives. One was 'mind-mapping', as developed by Tony Buzan in *Making the most of your mind*. The other was the method of exploring personal constructs through generating opposite poles developed in chapter 4, based on the *Personal construct theory* of George Kelly.

The use of these holistic tools of inquiry proved to be capable of generating considerable mental energy and enthusiasm for the task. Mind-mapping revealed that all practitioners saw their

Figure 9.6 Reasons for using patterns (after Paine 2005)

Field of work:	*Reason for using patterns*	*Opposite pole*
Acupuncture	Inclusive, insight	Narrow, blind
Architecture	Completeness	No form
Art lecturer, painter	Joining yin and yang	Disorder
Choreography	Healing the mind-body split	Fragmented individuals
Feldenchrais practice	Embodiment of integration	Fractured mind-body
Mathematics teacher	A trinity: two poles and a core	Apart from nature
Landscape ecology	Understanding complexity	Simplistic
Permaculture	The world is a pattern	Incompleteness, entropy
Psychoanalysis	Choose organising principles	One cause, a template, fixed
Physiotherapy	Body moves as one	Unrelated body signals
3 responses	Life	Death

Source: Paine, G. 2005, Wholes, Patterns and Sustainable Development: *Closing the Gap Between Concern and Action*. Unpublished PhD Thesis, University of Western Sydney, Richmond.

own use of patterns as important to their work, although these were not necessarily recognised as Alexandrian patterns. Each practitioner was asked why he or she used patterns in their work. They were then asked to position that reason against its opposite, in an application of Kelly's personal construct mapping (figure 9.6).

All the participants indicated that they purposefully sought to work with patterns. Paine identified five consistent aspects of pattern-making:

- a relationship between a pattern and a working with the whole;
- an association between pattern and 'life';
- new patterns invariably a result of a morphing of existing patterns;
- a similarity in the patterns developed by all holistic practitioners; and
- a tension between rigidity and fluidity in creating the pattern.

Pattern language was found to be acting as a lingua-franca between practitioner and client. The findings reinforced Alexander's proposition of the connection between the use of patterns and 'having life'. As the opposite pole to the idea of a pattern, three participants nominated death. Response from the physiotherapist: *you could have a messy pattern, but within that dis-organisation you can still have some sort of order.*" This is reminiscent of the interpretation of chaos theory in box 9.4.

In this chapter we have explored the

least widely acknowledged knowledge culture, holistic knowledge. We have found it alive and well, contributing along with the other knowledge cultures to any understanding of a whole. In the arts it is traditionally acknowledged as the dominant construction of knowledge, masking the other knowledge cultures, just as specialised knowledge masks the others in its particular fields. The genius of da Vinci in balancing empirical, aesthetic, social, ethical and introspective dimensions in his painting *The Annunciation* is conveyed by the power of his holistic vision. In physics, ecology and computing we have discovered that holistic thinking encourages us to look afresh at the world and events, and challenge perceived wisdom.

In pattern languages we have a tool that recognises holistic knowledge as a construction of knowledge in its own right. Further, it permits the construction of patterns by all interests. We therefore add pattern languages to the suite of synoptic tools for achieving collective social change.

Holistic knowledge lies at the heart of the suite of nested knowledges. A pattern language is therefore the final tool for collective thinking to be developed here, before we turn to considering collective knowledge as a mainstream activity. In the next and final chapter, we bring together the full suite of strategies which could produce the multi-faceted synthesis that is collective knowledge. We evaluate whether they are capable of achieving synergistic solutions to the wicked problems emerging in a disrupted human-environment relationship.

Box 9.5

TESTING: CREATING A COLLECTIVE IDEAL (3 HOURS)

Aim: to establish a shared holistic vision of their collective future for local citizens, community interests, local specialist advisors and responsible government agencies.

Participants: anyone with an interest in the region or place.

Facilitator: settle each group of seven in comfortable chairs in a circle close to a wall or whiteboard where you can pin up cards later. Spend 5-10 minutes discussing what it means to think 20 years into the future.

Key question for everyone to consider: In the *** region, what would people find if all the programs on social, economic and environmental sustainability in place now were working? (If there are people under 25 years of age, have them think forward into the unknown. If there are non-English speakers, be sure there is an interpreter). The group is in a quiet room with no interruptions. In a clear, slow, calm, deep voice, read the following script. Some people will have greater difficulty visualising than others. Pace the exercise so you leave plenty of time for everyone to visualise the images, and move slowly from one set of images to the next.

*We are now going to take a trip to the future, to an ideal *** region 20 years from now. This is not *** as it is today, nor is it *** as if everything had simply continued on as usual. This is*** as we would like it to be if all of our work toward a better future had been achieved. It is not pie in the sky, or an unrealistic Utopia, but what we can reasonably expect if we had the healthy, sustainable environment we wanted. Make yourselves comfortable. You may find it useful to close your eyes so you can more easily imagine the landscape.*

*I want you to come with me into our helicopter and first take it up high over ***, not *** now, but in 20 years' time when all our effort toward a better future has succeeded. Look at the landscape beneath you. What does it look like? What colours and shapes do you see? What time of year is it? Look out across the landscape at the cities, towns, buildings, spaces, farms, hills and valleys. What patterns do you see? What crops are growing? What stock is feeding? Where does water come from? How is the water distributed? How pure is the water? What is it used for? Where does the wastewater go? Where are the people? Who is in a family with whom? Can you see people and goods moving to and fro? How are they being moved? What sounds come up to you? What smells?*

Now choose a place you want to visit, to see what it's like in 20 years' time. Take the helicopter to that place. Descend slowly, looking around you as you come down at the patterns of buildings, of people of spaces. Look at the shapes and structures of the buildings. Look at the spaces between them. Who is moving in those spaces? As you get closer, listen to the sounds and smell the scents of the landscape.

Are there people around? Who is there? What ages are they? How do they react to

Box 9.5 (Cont'd)

TESTING: CREATING A COLLECTIVE IDEAL (3 HOURS)

you — and to each other? What activities are going on? Are there people working? Are there people relaxing? What else are they doing? Who is doing it? Now walk around and see what else there is.

Join some of the people as they move about. What are they doing? How do they look? Is this a good place to be? How are the people interacting with each other? Now join some other people. What is happening here and how does it feel?

Now imagine yourself in the open spaces. How is it different from today? What is happening in the neighbourhood and in the community? What are people happy about? What are they worried about? Imagine that it is lunchtime. Where do people go for lunch? What do they talk about? What sort of food are they eating? Where did it come from? What do they do after lunch?

As the day comes to an end, go home with someone. How do they get home? Where is home? How far are they travelling? Go with them in their neighbourhood. Remember, this is a well-managed environment in which all your programs and changes are working. How does the neighbourhood feel? Do you feel secure about the future? What important changes have happened recently?

Now go into a house. Who is there? Is it a family? Who is in the family? Are they all receiving an income? What do they do? What about the others? Are there children? Do they go to school?

Now climb back into your helicopter and take it once more out over the whole region, looking at the hills and valleys, rivers and towns. Bring it slowly back to [wherever you are] and come back to the present time, thinking about all you have seen. (Provide time here to return to the here and now). Now, before you forget what it was like, write down the 10 most noticeable things you saw, heard, smelled or touched.

Give the participants five minutes or so to write down the 10 things that stood out for them. If necessary, put some pressure on the participants to go to 10 items — the last two or three may be the most difficult, but often contain the most original ideas. Next, ask people to rank their 10 ideas or issues in order of importance to them.

There will always be someone who is very practical, or not at all visual, who has trouble doing the exercise; that's fine. They should be made to feel comfortable and asked to complete the 10 ideas from guess-work or their own experience rather than guided imagery.

Form groups of seven to 10 people around the tables, far enough apart not to disturb each other. Ask each person to write on three cards, in five words or fewer and in big letters people can read across the room, the three most important ideas among their 10. Each person is asked to stick his or her most important card of the three on the whiteboard or wall themselves (note: each participant does this, not the

Box 9.5 (Cont'd)

TESTING: CREATING A COLLECTIVE IDEAL (3 HOURS)

facilitator) and then read out the idea. Others may ask questions of clarification, but they may not offer comments or criticism. Each person should either put his or her card in a group with the others or start a new category. Don't let any of the participants be critical of anyone else's issues, but encourage each person to describe what their card means to them, and people to ask sympathetically what it means to them.

Then take a second round of cards, then the third. No criticism of any contribution is allowed. Short questions of clarification are permitted. Everyone puts their cards wherever they wish, and no one can move another person's card without permission. When the cards have been collected into a minimum number of strands (probably about three, perhaps five or seven) each strand should be given an imaginative name or title encapsulating the central idea. This then gives the central themes for a sustainable and healthy environment in ***. In a plenary session of all the groups involved, the themes are then reconciled into a minimum number of sustainability goals for the region (this is surprisingly easy, due to the process participants have shared).

After the themes have been agreed, take another half-hour for everyone to suggest how they would know whether the goals have been achieved. From the cluster of observations around each goal, select one.

Outcome: 10-20 primary sustainability goals and their indicators recognisable by all decision-making interests in the region.

Postscript: The strength of the visioning method is that it is inclusive, not competitive: everyone's key goals can find a place. Negotiations can go on 'beyond win-win' to a third possibility that increases everyone's outcomes. The essential basis is that everyone undertaking the exercise belongs to or has an active interest.

Source: Adapted from Healthy Cities visioning process, designed by Trevor Hancock, 1986

10 A just, secure & sustainable future

"Certainty is born of a complete knowledge of all the parts, which, when combined, compose the totality of the things which ought to be known."

Leonardo da Vinci, *Notebooks*, 1489

Three bricklayers were asked what they were doing. The first answered that he was laying bricks, the second that he was constructing a wall and the third that he was building a cathedral. A fourth could have answered that they were part of a team laying bricks for a wall of a beautiful cathedral. This last answer may be longer, but has the virtue of acknowledging ideals, facts, ideas, and actions. The metaphor of the fourth bricklayer is one way of symbolising the goal of the journey we have been taking. Another is Thomas Berry's description of the goal of a collective movement towards sustainability as the 'great work' of our time. Yet another is Leonardo da Vinci's warning at the head of this chapter. All of these celebrate a commitment to grasping the whole and the parts at the one time.

What should be:
the fourth bricklayer

The search for ways of achieving synthesis began with a set of big questions.

- How would adopting the goal of synthesis change what we are doing now?
- Can the changes be absorbed into present ways of thinking?
- Could the changes amount to a new renaissance?
- If synthesis becomes central to thinking and action, is that sufficient in itself to guide us in harmony through the rest of the century to a more sustainable future?

It is now time to review the possible answers. The evidence has been drawn in part from the ten field studies of synthesis in action, in part from examples of Leonardo's life work and in part through drawing on the work of Renaissance-like synthesists of our own time. It seems that the capacity for making collective decisions is here. It is not at all clear whether it will remain marginalised, short-term and alternative, or whether it can become a core part of the rules and practices of society.

Taking the second question first, if working with synthesis were to become the rule, there would inevitably be a social transformation. This was true of the prototype, the first Renaissance. In less than a century, there was a giant leap forward in individual creativity, public governance, citizen responsibility and financial management. This was the time that produced da Vinci, Machiavelli and Dante among others. Worker's guilds gave community support. The first public bank was backed by the wealth of the Medici family. Freedom of religious belief was incorporated in the city charter of Urbino. These changes generated a synergy that ran throughout the fabric of a thriving society.

Thirteen generations later, there are signs that the latest generation may be taking up the Renaissance bequest. Individuals are accepting the need to develop their full portfolio of talents. Learning from difference in multi-skilled teams has become standard practice for many organisations. Communities are taking up the advocacy role of civil society. Specialists are incorporating personal, local and strategic evidence in common sense inquiries. A focus on holistic rather than partial goals can be seen in the spread of the concept of sustainability.

Box 10.1 contains the long list of the Renaissance thinkers who have provided the foundation for the argument for this potential change. Between them they include expertise in the full set of knowledge cultures of the Western decision-making system. Their primary concerns are individual knowledge (Berlin, Bohm, Kelly and Kolb); community knowledge (Alexander, Duchamp, Midgely, Mumford and Shiva); their own specialisation (Gould, Lovelock, Toulmin and Waring); organisational knowledge (Bawden, Falk and Saul) and the search for the essence (Berry, Koestler, Smuts, de Chardin and White), respectively. Taken together they bring various combinations of science, arts, humanities, religion, technology, politics and commonsense. Their connecting principle, however, is that they themselves access each of the knowledge cultures identified in chapter 3, and try to answer the full set of commonsense questions established in chapter 7.

Each of these synthesists has contributed to the portfolio of tools that bring the various decision-making systems together. Their work provides the components for a culture of synthesis. The synthesis itself has been achieved through an action learning framework, decisions-into-practice, d4-p4 for short. In each of the five previous chapters, the stages of the collective decisions-into-practice spiral – *what should be, what is, what could be*, and *what can be* – have formed the framework for the chapter. Each stage has been presented here with its own suite of tools that enable synthesis in thought and action. The summaries of three of those stages in figure 10.1 allow us to answer the first of the questions with which we began: how would a commitment to synthesis change what we do now? The answer is that where there is a commitment to synthesis at every stage of the construction of knowledge, there is also a shared (but not necessarily uniform) construction of reality and ability to act.

All the capacities included under the heading *what should be* in figure 10.1 from searching synthetically to traveling hopefully are already part of the human repertoire. However, they are far from being included in the standard Western constructions of knowledge. Applying Smithson's anatomy of ignorance to figure 10.1, each of the synthesis-based ways of thinking under the heading of *what should be in the first column* are frequently rejected as errors of judgment. Or else they may be dismissed as irrelevant. It is therefore not to be expected that the ways of developing collective knowledge listed in figure 10.1 will be found in many higher education courses or selection criteria for employment.

However, they are beginning to be found in personal, organisational and community development programs, and in courses addressing wicked prob-

Box 10.1
LIBRARY OF RENAISSANCE THINKERS
OF THE 20TH-21ST CENTURIES

Synthesists and their expert fields:	*Contribution to collective thinking:*
Alexander, Christopher, architect	A pattern language
Bateson, Gregory, anthropologist	An ecology of mind
Bawden, Richard, systems agriculture	Holocentric organisation
Berlin, Isaiah, philosopher	Unifying power of ideas
Berry, Thomas, philosopher and priest	A moment of grace in which to act
Bohm, David, physicist and thinker	Critical dialogue and learning from difference
Brin, S. and Page, L. software designers	Relational search engine Google
Duchamp, Michel, artist	Art as an agent of social change
Falk, Richard, international relations	Globalisation from above and from below
Foucault, Michel, social biologist	Unity of power and knowledge
Kelly, George, engineer and psychologist	Personal construct mapping
Koestler, Arthur, writer	The act of creation
Kolb, David, educational psychologist	The experiential learning spiral
Lovelock, James, analytic chemist	The integrity of the planet Gaia
Midgely, Gerald, critical systems designer	Systemic intervention methods
Mumford, Charles, writer	The city as symbols
Saul, John Ralston, sociologist	Synthesis between ethics and reason
Shiva, Vandhana, nuclear physicist	Indigenous rights to traditional knowledge
Smuts, Johann, soldier and politician	The organic, systemic, holistic universe
Teilhard de Chardin, archeologist	Evolution of collective thought
Toulmin, Stephen, philosopher	The analysis-synthesis pendulum
Waring, Marilyn, economist and politician	Integrity in national accounting systems
Wright-Mackinney, Judith, poet-activist	The Human Pattern

Sources: see Bibliography

lems such as sustainability. Would-be synthesists who are practicing the principles outlined in column 1, figure 10.1 have the tools available to achieve the full set of outcomes listed under *what can be* in column 3. These range from searching synthetically for a synergistic system to traveling hopefully towards a new Renaissance. The central column of the table lists the parameters that determine the components of each component of synthesis.

The shift from the individual to the collective is symbolised by two forms of birthday greeting. To mark a birthday in

Figure 10.1 Collective principles, parameters and products: a journey of synthesis

What should be: principles *Ways of synthesis*	**What is: parameters** *Parts for a synthesis*	**What can be: products** *Tools for synthesis*
Chapter:		
1. Searching synthetically	Cores and boundaries	A synergistic system
2. Working synoptically	A jigsaw puzzle	A creative collage
3. Knowing collectively	Five knowledge cultures	Nested knowledges
4. Acting synergistically	Debate and dialogue	Networked knowledges
5. Reflecting introspectively	Range of learning styles	A collective mind
6. Communing powerfully	Diverse community symbols	Shared symbols of place
7. Specialising systemically	Relevant set of disciplines	Set of commonsense questions
8. Organising ethically	Internal reference	Holocentric management
9. Talking holistically	Creative arts	Pattern language
10. Travelling hopefully	Utopian goals	A new Renaissance

the individualist Western tradition, we sing 'Happy birthday, dear John/Mary, happy birthday to you' followed by three cheers for the important birthday person. The goal of Montessori education is to celebrate the whole child in the whole of their social and natural environment. Their song for a 10th birthday is 'Ten times round the sun, John/Mary has been 10 times round the sun', with the birthday child one link in a dancing circle of children. Throughout our journey, we have found that each of the end-points of *what can be* in figure 10.1 brings with it a changed pattern of behaviour, as well as a change in ways of knowing.

Current practice segregates ideals, facts, ideas and actions on an issue. As Leonardo warns in the quotation above, when we do this the whole has vanished from our grasp before we have even begun. While accepting that it is not possible to fully know the whole of any issue, this does not prevent synthesists from reflecting on the whole, before distilling its essence. Throughout Western society, there are many fourth bricklayers who manage this task – they can be recognised under many headings, as synthesists, polymaths, change agents, integrators, gatekeepers, tussock jumpers, organisational termites and the people in the cracks.

The evidence presented here confirms that a tradition of collective thinking and constructive dialogue is developing among a small band of committed people. There are the equivalents of the fourth bricklayer in many collaborative teams. Is this sufficient to establish a sixth knowledge culture in its own right? If it is, then a paradox immediately arises. One the one hand, a collective knowledge culture accepts the value of open learning, uncertainty and ambiguity. On the other hand, it forms yet another circumscribed way of constructing knowledge. The next task is to revisit and evaluate whether the synthetic processes listed here are capable of resolving that paradox.

What is:
a changing knowledge tradition

"*The part always has a tendency to unite with its whole in order to escape from its own imperfection.*" Leonardo da Vinci, *Notebooks*, p281, ref 44.

The analytic tradition has divided the ways of knowing into the parts listed under 'what is' in figure 10.1. The field studies described in chapter 4 confirm that the hierarchical knowledge tradition ensures that for each stage of the decisions-into-practice learning cycle one aspect will be given greater emphasis than the others. Boundaries will receive more consideration than cores. Linear thinking is formally rewarded more often than lateral thinking. Synthesis has been expected to take the form of consensus, leading to the lowest common denominator. Under these conditions, the parts remain impediments rather than contributors to synthesis.

In a synthetic process, on the other hand, starting with the whole means the parts can be considered afresh, in relation to each issue. Each part is regarded as making an equally valuable contribution to the whole. Thus, each knowledge culture is respected and fully included. In the previous chapters, the sections on 'what can be' have ensured that each of the knowledge cultures considers the evidence from each of the others. The tools of synthesis have been based on three generic strategies: thinking synoptically, knowing collectively and acting synergistically.

To think synoptically is to bring together a range of understandings on one issue, leading not to consensus but to a deeper understanding of the whole. Niels Bohr's idea of complementarity – in which different states, such as light as particles and as waves, co-exist – has already been quoted as an example of synoptic thinking. The most productive structure for synthesis at each stage has proved to be a network. The knowledge cultures then interact in a web-like pattern, forming the dynamic whole-of-community systems described in chapters 6, 7 and 8 (figure 10.8 below).

In a synergistic synthesis, the liaisons among the knowledge cultures at each learning stage allow them to learn from each other, rather than becoming competitive and defensive. Programs using the decisions-into-action learning spiral have included doctoral theses, whole-of-community monitoring, a textbook to change professional practice, whole-of-organisation change, and a pattern language. Some valuable lessons on uniting their parallel worlds emerged for each of the knowledge cultures.

Individuals and their collective genius

As an antidote to the divisions in personal learning styles, collective thinking requires that each person diagnose their own preferred learning style and knowledge culture, as in the exercise in chapter 5. They may then develop the styles and cultures that they have been so far neglecting, or they may choose to work in a team where all are represented. Such collective thinking using all abilities, as individuals and in groups, is found among Nobel Prize winners and ordinary citizens alike.

The individual's personal capacity for synthesis has been identified as the 'glue' that holds a more general capacity for synthesis together. Since all knowledge has begun in an individual human head, any change in ways of knowing will have to take place there too. The power of that personal synthesis is demonstrated in the individual gestalt, as in the 'Christ in the Snow' figure, and in each individual's symbols with which they identify their community.

Whole-of-community engagement

Shared symbols that hold significant meaning for everyone in a community legitimise each community's local knowledge. The symbols offer community members a shared language through which to make sense of their stories. Mumford describes every long-lasting community as a synthesis between the power exerted by palace, temple and granary; social, ethical and economic power. Once the symbols are shared across all the citizens, rich, diverse, whole-of-community visions can emerge.

The different knowledge cultures combine synergistically where there is a shared symbol for a preferred future. Berlin preserves the remains of the Berlin Wall. Bob Geldof promises an annual LiveAid. The collective response to social and environmental disruption has been given the label of sustainability. Cultural anthropologist Greg Dening has developed islands and beaches as symbols for whole-of-community change. A community is a self-organised island, with its surrounding beach as the zone of its interaction with the rest of the world. People approaching the community may cross the beach as missionaries, as traders, or as conquerors, or they may remain on the beach as beachcombers, part of visitor and community alike.

Transdisciplinary inquiry and commonsense

Every specialisation, whether in the arts and sciences, or the trades and professions, conducts their work within a specific framework or paradigm. It is the strength of the boundaries between the specialisations and the rigour of the definition of their core, that enables specialists to arrive at their detailed knowledge. However, this in turn makes cross-specialisation hard to achieve, and including all the knowledge cultures almost impossible without remedial action.

One way of adjusting for the separation among the knowledge cultures is to consider the answers to a set of commonsense questions, including the empirical, social, ethical, aesthetic, introspective and reflective. The more familiar categories of economic and political are subsumed under the social, since they are social constructs that vary according to the society. The knowledge cultures each have their particular question they are accustomed to answer. However, each knowledge culture already examines the evidence to answer all of the questions, but implicitly. In a transdisciplinary inquiry incorporating all the knowledges these questions are asked explicitly. The Local Sustainability Project field studies and current Renaissance thinkers have identified a range of ways to bring specialised inquiry into dialogue with the other knowledge cultures in answering commonsense questions. The result is the eclectic suite of methods that forms transdisciplinary inquiry.

Holocentric organisational structures

In a knowledge synthesis, one of the most difficult hurdles to overcome is the idea that organisational power is necessarily based on principles of divide and rule. The standard organisational structure is hierarchical and compartmentalised, with little consideration of connections between the parts. In practice, any organisation functioning within a society is at the meeting point of two sets of principles, whether or not these are made explicit. Each set of principles, liberty and equality, and community and efficiency, contains its own tension, which can be increased or reconciled.

A collective approach seeks to resolve the tension within the two sets of principles synergistically, that is, in a way that they can contribute to each other and to a different outcome. A holocentric management structure bringing them all together can turn either inwards towards focusing on the organisation itself, or outwards towards a democratic state, or both. The tensions can be resolved in practice through the way in which organisational structures actually operate. A review of 16 organisational types reveals that any form of organisation can be either holocentric or hierarchical, although some structures make it easier to be holocentric than others. Examples of holocentric management structures observed in the field studies include a whole-of -community action web, and an advocacy group for local sustainability in the local government sector (chapter 9).

Holistic pattern languages

Holistic thinking discovers the essence, the core understanding, of any idea. As such, it is an inherent part of any decision, individual and collective. A fresh holistic focus is by definition a driver of change. Pattern languages for water, aging and civic communication open up different ways of approaching their themes (figures 9.4, 9.5). A fresh twist to the existing social concept of money as a token of a share in the market economy is presented by performance artist, John Reid. His memorable design for a series of banknotes in which the capital being drawn on is the natural resources of the planet can be found in box 10.2. The bank notes are exchanged for standard bank notes and the money allocated to sustainable environment projects.

In making a collective decision, there must be some shared understanding of a mutually satisfactory outcome. It is important to note that it is not intended to be a single monolithic value, nor a single authoritarian direction, but an agreed purpose towards a better future. Diversity of interests enhances dialogue and generates synergy when-

Box 10. 2

CHANGING THE SYMBOLIC MEANING OF MONEY

(JOHN REID, PERFORMANCE ARTIST, 2002)

The National Environment Bank (NEB) was conceived by visual artist John Reid as a contribution to an exhibition entitled Factor of Ten – A Future Worth Having in Canberra, Australia, in 2002. The exhibition canvassed the need for a 10-fold reduction in human energy and material consumption to achieve a sustainable use of the Earth's resources. NEB branches are operating in nine regional art galleries. Branches consist of a mint, money exchange, investment bureau and a collection of promotional, framed photographic prints of Australian riverbanks. The NEB Mint prints commercial paper in a range of denominations – \$10, \$20, \$50, \$100, \$200 and \$500. The banknotes feature photographs of Australian riverbanks that are crucial repositories of plant genetic material and sites of cultural heritage.

Artist's statement: The NEB Investment Bureau invests the revenue raised in sustainable energy alternatives. The bank notes are a link between the aesthetic power of visual fine arts to inform and motivate and the enabling power of money. Add to this the veracity of 'straight' photography and the bank notes become fighting documents with the potential to challenge lethargy and unethical behaviour by governments and their agencies. Each bank note sold stands as a quantitative measure of individual discontent with the collective disregard for the wellbeing of the planet.

Reference: www.neb.com.au [accessed 25.07.05]

ever they are brought together under the umbrella of a holistic idea. 'Cool communities' as the title for a project to reduce global warming is one such idea, a pattern language for civic communication is another.

Is there a collective construction of knowledge?

We are now in a position to give more conclusive answers to the first two of our original questions. For the first question - how will valuing synthesis change what we do now? - we have found that the use of the tools for synthesis changes the construction of knowledge and, therefore, the shared construction of reality. This requires the acceptance of a considerable degree of personal and professional change.

For the second question - what is the extent of the social change required? - the answer is that we do not yet know. We do know that moving to synthesis is a tool for social change and involves social change. It is not a revolution, in that the move values existing abilities and processes. It is not evolution, in that it is not a continuum but a break with the past. Whether it is long-term change to the extent of a social transformation remains to be seen.

Answering the third question - are the tools capable of bringing about moves towards a just, secure and sustainable global future? - is much more difficult. In some circles it is dismissed as hopelessly utopian even to pose the question. However, the general consensus of experts, organisations and concerned communities is that a collective way of thinking is essential in bringing progress towards the utopian ideal of sustainability.

Reviewing the essential characteristics of collective knowledge in figure 10.1 it is comparatively straightforward to demonstrate that the abilities listed in figure 10.2 can service the new paradigm in the Kuhnian sense. We find that they match the criteria for a fresh framework for knowledge construction. There is a distinctive form of synoptic content in nested knowledges; specific synthesis-based tools for thought in concept mapping and discourse; an eclectic and synergistic mode of inquiry in networking the knowledges; relationship-based tests for truth in answering commonsense questions, a synergistic exploratory framework in the social learning spiral d4-p4; a particular type of question directed towards the wicked problem of a sustainable humane future, and many individual synthesists as role models. Therefore, constructing knowledge collectively has the characteristics of a fully distinct knowledge paradigm, in theory and in some professional practice. It is less certain that the characteristics are yet sufficiently accepted to amount to a new knowledge culture in practice.

What could be:

tools for a collective knowledge culture

"The senses are of the earth; reason stands apart from them in contemplation." Leonardo da Vinci, *Notebooks*, p6, ref 10, Triv 33.r

Figure 10.2 A collective knowledge framework (after Kuhn 1962, 1972)

Dimensions	Collective knowledge framework
Content	Synoptic: compiled from individual, community, specialised, organisational, holistic and collective interpretations
Tools for thought	Synergistic: Chaos theory, rules of dialogue, construct mapping, wicked problems, paradox, symbols and commonsense questions
Mode of inquiry	Synergistic: networking in collaborative action research; asking the full set of commonsense questions: empirical, social, aesthetic, ethical, introspective and reflective
Tests for truth	Synoptic: Dialogue and critical loyalty among the contributors Answers to commonsense questions validated by their appropriate criteria for truth
Explanatory framework	Synergistic: Decisions-into-action open learning spiral
Type of question	Holistic: reflective questions on the whole
Role models	Synthesists: *Leonardo da Vinci, Teilhard de Chardin, Vandhana Shiva, Thomas Berry, Lewis Mumford, Fritjof Capra*

A culture is much more far-reaching than a paradigm. While a paradigm is a lens for examining reality, a culture is a coherent system made up of the ways in which people live, talk to one another, interpret events and construct a shared reality. In considering the requirements for synthesis, several of these aspects of a culture have been discovered to already exist. It is too early to assume, however, that we will be able to achieve a change comprehensive enough, or interwoven enough, to establish a thriving culture, as there was during the 15th century Renaissance.

Studies of synthesis in action have revealed some connecting themes that open up at least the possibility of a synthesis-based knowledge in the future. Setting the scene for a possible collective knowledge culture are the open systems for global communication in cyberspace, complexity and chaos theory as a way of interpreting reality, and nested and networked systems for constructing new knowledge.

Cyberspace: the noosphere?

Each of the knowledge cultures has developed modes of communication through the internet. That the internet was preserved as common ground in the first place was due to collaboration among a few individuals working in information technology and at the universities. Continuing efforts to restrict the open communication through copyright, hidden codes and constraining lawsuits have been countered by creative computer professionals. These have designed open codes everyone could use freely; improved on the copyright commercial codes; opened watertight compartments; and called on existing protections of freedom of speech. The rest of us can be grateful for their constant efforts to preserve the open system.

Maintaining cyberspace as a public communication sphere has meant that it remains a common resource for the practice of synthesis. Systems of open access networks, internet passports and community networking are replacing the one-way stream of information with restricted access.

At the beginning of this discussion we considered Hesse's classic story of the glass bead game, whose aim was to find a method of bringing together all the knowledge in the world. In 2007, the most powerful computing designs are being directed towards placing all the products of human thought on the Internet. The wikipedia is a reliable interactive storage space for all information and the gutenberg project is placing all literature on open access, both advances towards this utopian goal. The capacity to make use of these tools for collective thought is being advanced by the complex computer games that children learn to play almost from infancy. In box 10.3 is an announcement of an internet-based program that reads like nothing else but a real-life glass bead game.

Box 10.3
DIGITAL GAMING A WINNER FOR 21ST CENTURY SCIENCE

Says Jim Brazell, VentureRAMP principal. "In the next 5 years, we will see our concept of cyberspace transformed from web pages and bookmarks to network worlds and new forms of human communication, work and collaboration based on game-like technologies." Gaming: A Technology Forecast delves into these ideas and illustrates how gaming is linked to 21st century science and an emerging academic revolution called transdisciplinarity. The game report may be downloaded free online at http://www.forecasting.tstc.edu/forecasts/digitalgames/.

Source: Brazil, J. email to Pattern Languages Yahoo Group

Commercial software designers of Geographic Information Systems have software that can project future options from present information. There are striking parallels between the directions in which the use of cyberspace is taking the capacity of the human mind, and the predictions of the early authority on human evolution, Teilhard de Chardin. Half a century ago, before the idea of cyberspace existed, he wrote that just as tool use had developed our biological brain, responding to social complexity was developing a collective human brain, the noosphere. He predicted the shared centre of the noosphere, *point omega*, as the next evolutionary leap. With the continual emergence of fresh collective ways of using cyberspace, this idea does not seem so unrealistic.

Chaos theory: whole in the parts and the parts in the whole

In chapter 9, the seven dimensions of chaos theory have been identified as creativity, connectivity, self-organisation, intermittency, fractal space, fractal time and holism. It is worth noting that these dimensions resemble Christopher Alexander's requirements for centres in a pattern for liveable social-environmental design. The dimensions are linked to the idea that small changes can have vast-reaching, unintended and unpredictable ramifying effects. Leonardo's proposition: "*The earth is moved from its position by the weight of a little bird alighting upon it*" – in his Notebooks in 1489 is not so very different from the classic 1979 paper by Edward Lorenz: "*Does the flap of a butterfly's wing in Brazil set off a tornado in Texas?*"

One of the discoveries arising from chaos theory is the ubiquity of fractals throughout the biological and physical worlds. Fractal is the word coined for the units of patterns that contain little copies of themselves. Buried deep within the original pattern, they generate infinite detail by repeating themselves. The more you zoom in on a fractal, the more detail you get, in a coastline or a weather pattern. Box 10.4 contains an example of a fractal formula that determines the patterns of snowflakes, coastlines, human lungs, lightning, seashells and brindled cows. Poets such as Hopkins and Whitman have long recognised this universality. The quote from Leonardo in box 10.4 confirms that he predicted the existence of these connections, although it was too early to identify the mechanisms.

Networking: small worlds

Earlier, we adopted Donella Meadow's definition of networking as "*a web of connections among equals...What holds it together is the understanding that some tasks can be accomplished together that could never be accomplished separately*". In chapter 3, collective thinking was represented as networking between the knowledge cultures. In chapter 6, work on whole-of-community change resulted in a change management web. In chapter 7, a set of commonsense questions that included all the knowledge cultures created a web of collective inquiry. In chapter 8, linked holistic patterns took the shape of a web, forming a pattern language. Networking has proved to be the essential process that underlies collective thinking and action.

Communication webs have certain characteristics in common. In the early 1990s American social psychologist Stanley Milgram found that there were no more than six degrees of separation between any one human being and another in the United States, with its (then) 300 million population. A burst of work in the early 2000s found that webs were a ubiquitous form of organisation throughout the social and physical worlds. Confirming Milgram, between six to nine degrees of separation is a characteristic of them all, from friendships in a city to the separation of a species in evolution, and connections made on the internet (figure 10.3)

Albert-Laszlo Barabasi in *Linked. The*

Figure 10.3 Communication pattern found in foodwebs, computer searches, citations etc (after Barabazi 2002)

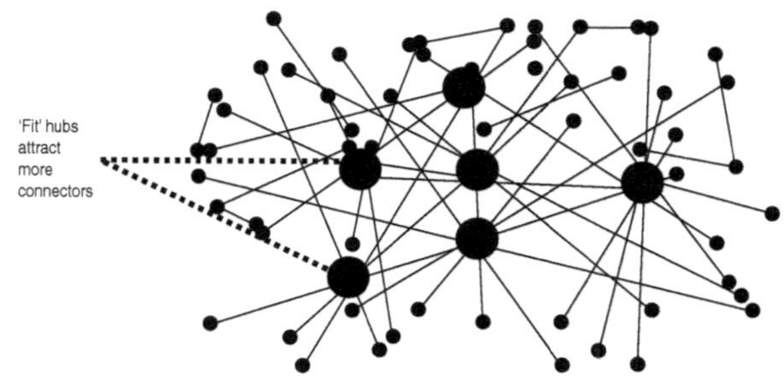

New Science of Networks describes how network C (figure 10.3) develops spontaneously throughout social and natural systems, wherever there is an opportunity for innovation. Examples are as varied as the social support systems that create social capital, and the food webs of ecosystems. Such networks or distributed systems are truly 'small worlds'. They form whenever a fresh field emerges, be it a new community, profession or organisation. The networks cluster around 6 to 9 principal hubs, which he describes as 'fit' hubs from their power to attract other hubs. As the number of links in the field increase, the network moves from figure 10.3 to one of the networks in figure 10.4.

In his book *Nexus. Small Worlds and the Groundbreaking Science of Networks*, science journalist Mark Buchanan reviews work by IBM computer scientist Paul Baran. Baran identified standard communication patterns in a wide range of organisations. His computer modelling found network C with its cross-linked centres, more robust than network B's linked cores, and far more robust in response to external change, and responsive to internal change, than Network A with its single centre.

Relating to people and the environment through networks, rather than linear logic or closed cycle problem solving, is found in all aspects of human activity and fields of human inquiry. The outcome may be called harmony, balance, fitness, beauty or creativity, according to the field of the speaker. Networks form the fabric of most complex systems, from a community to a nation; from a cell to an ecosystem. They have the potential to help us form a collective understanding of a complex, dynamic, interconnected world.

What can be:
a sixth knowledge culture

It seems a case can be made for a sixth knowledge culture within the Western tradition. Many instances of the collective construction of reality already exist, and indeed have always existed. One only has to be present at a standing ovation for a singer such as

BOX 10.4 THE PARTS IN THE WHOLE, AND THE WHOLE IN THE PARTS: FRACTALS IN POETRY, MATHEMATICS, BIOLOGY, PHYSICS AND MEDICINE.

Snowflakes and coastlines

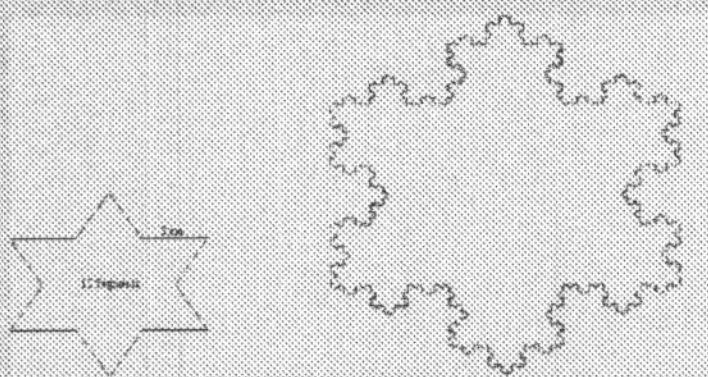

$$N=\left(\frac{P}{p}\right)^{d} \quad \text{-or-} \quad d=\frac{\log N}{\log (P/p)}$$

Formula for a fractal: Koch's curves

Pied beauty:
GLORY be to God for dappled things,

For skies of couple-colour as a brindled cow,
For rose-moles all in stipple upon trout that swim;
Fresh-firecoal chestnut-falls, finches' wings;
Landscape plotted and pieced, fold, fallow and plough,
And all trades, their gear and tackle and trim.
All things counter, original, spare, strange,
Whatever is fickle, freckled (who knows how?)
With swift, slow; sweet, sour; adazzle, dim.
He fathers-forth whose beauty is past change;
Praise him.
Gerard Manley Hopkins 1860

Medical diagnosis: human lungs,
with and without asthma

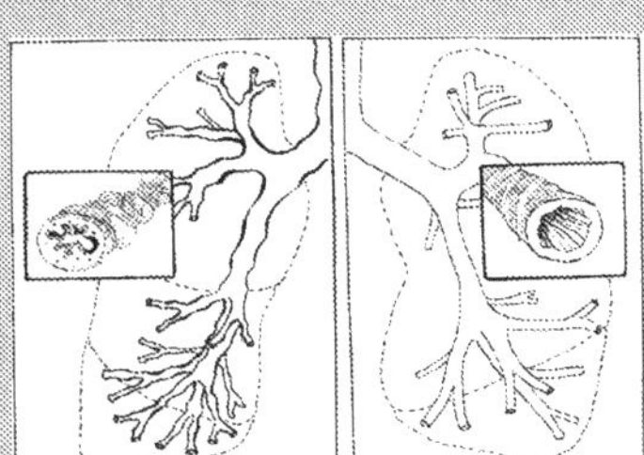

Weather forecasts: Mandelbrot 'lightning' set

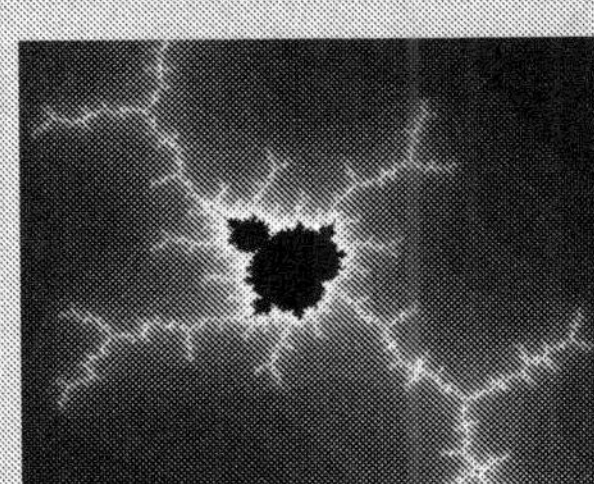

Every body placed in the luminous air spreads itself out in circles and fills the surrounding parts with an infinite number of images of itself, and appears all in all and all in each part.
Leonardo da Vinci 1498 Notebooks p38 ref 50

Figure 10.4 Centralised, decentralised and distributed network patterns (after Baran 1967)

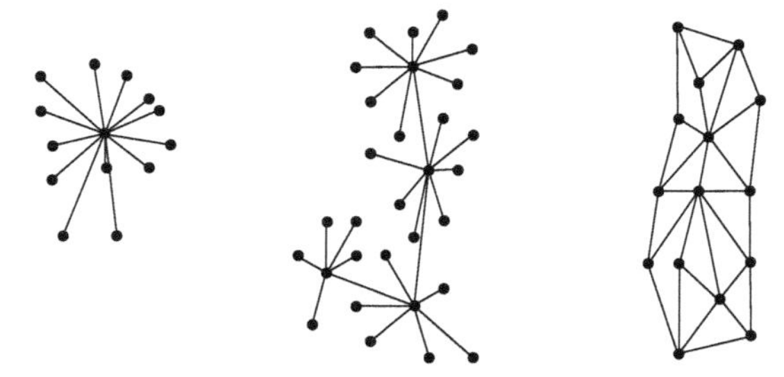

Cat Stephens or at a convivial community event to recognise that collective learning has taken place. Collective learning is a mark of many traditional societies, and of certain times of history such as the Renaissance. The Local Sustainability Project field studies have confirmed that uniting the divided knowledge cultures of the Western tradition in the pursuit of sustainability has provided stimulus for change along many dimensions. A collective knowledge culture follows the set of principles governing a guided synthesis (10.5).

We now come once more to the fourth question with which we began. If synthesis is a matter of social transformation, is that social change sufficient in itself to ensure movement towards a humane, sustainable future? Leaders in all the knowledge cultures have confirmed that collective thinking is a necessary condition for a move towards sustainability. In the field studies of synthesis as a core practice within each of the knowledge cultures, the goal of sustainability was the common driver for them all. Of the 25 proponents of synthesis in figure 10.1, 20 developed their ideas in the context of a humane, sustainable future. However, the present inquiry has focused on managing a change in the construction of knowledge. We have not addressed managing changes in advances in technology, nor concerted changes in overall social behaviour. We are therefore not in a position to decide whether the change in knowledge management is sufficient by itself.

A further common thread between the people listed in figure 10.1 is that, whatever their theme, they have sought to answer all six of a possible set of commonsense questions: empirical, social, ethical, aesthetic, introspective and reflective. Although this may appear straightforward, in practice it has proved not as easy as might have been expected.

In an analytic culture, there is a tendency to take the part for the whole. One form of question is addressed in isolation from the others. Introspective and reflective evidence is not taken seriously in relation to the rest. Ethical and aesthetic matters tend to be subsumed within social matters. Economic and

Box 10.5
PRINCIPLES FOR THE COLLECTIVE CONSTRUCTION OF KNOWLEDGE

Working collectively:

Keep the whole in mind, *reflectively*

Hear all community voices, *sympathetically*

Ask commonsense questions, *synoptically*

Arrange strong strategic alliances, *purposively*

Search for the heart of the matter, *holistically*

political questions are addressed as if they are separate from the society that has generated them. We have pointed out that this is a flawed logic, since economics and politics are inherently social, and conform to social rules. Ethics and aesthetics refer to different dimensions of experience from the social and the empirical.

As leading contributors to the art, science and politics of collective thinking and action, there is inspiration in the work of the following synthesists as they answer all the commonsense questions, but concentrate on one.

Introspective questions: Isaiah Berlin

The 20th-century reviewer of the history of ideas, Isaiah Berlin, never wavered from his conviction that the human capacity for interpreting reality rested on the strong drive in every individual to understand the whole. He clarified the considerable difference between regarding the world as a set of different parts (like a fox) and as a unified whole (like a hedgehog). He explored the difference between experiencing the world as a strategist, and through direct experience. His examples are Tolstoy's description of a battle from the experience of the general and of the ground troops. Each has totally different expectations and experiences of the same battle – both are valid. Those expectations and experiences are part of all our existence and are, to a great extent, self-fulfilling prophecies that shape our individual lives and society as a whole.

Social questions: Michel Foucault

The French social philosopher Michel Foucault points out in *The archeology of knowledge* that in ancient Greece, power was the property of a god, during the Renaissance, of a prince. Since that time, power has come to flow haphazardly through the interactions between organisations, bureaucracies, media, citizens and civil society. The relationship remains fluid, dependent on changing alliances and events. As a result, Foucault has suggested that everyone and no-one is in charge of the world. Chaos theory operates in the social as well as the physical world.

It is true that no single entity holds the absolute power of a Renaissance pope or of a family such as the Florentine Medici. Anyone aspiring to power now claims to speak for 'all the people'. Real power now functions silently, not through edicts, written rules or mythology, but by the unspoken permission of 'all the people'. Boundaries are set up which allow those who do not agree with the flow of power to be classified as 'the other' – including terrorists, overactive children, gays and the homeless. They are thus placed outside 'normal' society. But they have to agree to be controlled by society, for the system to keep functioning. Once outside, they have increased leverage for change. This is where the organisational termites and boundary-riders come in.

Ethical questions: Thomas Berry

We began this journey recognising the force of the theologian Thomas Berry's point that our moment in time is a moment of grace in which we can choose our future. Those alive now hold the power, if they so decide, to send the social and environmental condition of the planet in any direction. The collective force of the writers and institutions that we have been reviewing confirms his proposition. What other advice has Berry that needs to be heard?

His hope for the future rests on the power of individuals to make decisions for themselves, within a moral society that mediates for the good of all. He asks for recognition of the unity of the living and non-living components of Earth. Berry argues that a viable human future depends on a human ethic that includes individuals accepting responsibility for their own decisions, recognition of a higher good beyond the reach of any single individual, and respect for the integrity between living and non-living elements of the planet.

Aesthetic questions: Gregory Bateson

Gregory Bateson, whose work combines biology, anthropology, psychology and philosophy, was the son of one famous biologist and husband of another, anthropologist Margaret Mead. His idea of an ecology of mind is grounded in the dynamic patterns of interconnection among all living organisms. Outside the Western tradition that draws separately on art, poetry, architecture and literature, and which remains individualistic, Bateson's ecological aesthetics are essentially collective and connected.

For Bateson, the sensation of beauty or ugliness is an outcome of co-evolution. The reaction is a patterning developed from feedback loops from experience and environment over evolutionary time. His most memorable phrase is 'search for the pattern which connects'. Symmetries, contrast, colour and other types of branching pattern that we see in flowers, mathematics and music in their various ways relate to our own nervous system. Our resultant human sense of form and symmetry resonates with the presence of connecting patterns wherever we may find them. The stimuli for a creative leap are essentially holistic. Symbols, sounds and experience carry the essence of the whole pattern.

Empirical questions: Teilhard de Chardin

The French Jesuit priest, Teilhard de Chardin, was the paleontologist who discovered the evidence of human evolution in China in the middle of the 20th century. He was one of the first scientists to recognise the major shift involved in arriving at a collective understanding of reality: *"Man (sic) is no longer, as we would formerly conceive him, the immobile centre of an already completed world – on the other hand from now on he tends, in our experience, to represent the very leading shoot of consciousness, of a universe that is in the process, simultaneously, of material 'complexification' and psychic 'interiorisation', both processes continually accelerating."* (Tielhard de Chardin, *Man's place in nature*, p13)

He concludes: "*It is a vision whose impact should strike our minds with such force as to raise to a higher level, or even to revolutionise, our philosophy of existence.*" As a scientist Teilhard observes that the average life of a species is about 50 million years, so that there is plenty of time to achieve his predicted development. As a humanist, he admits that the same pressures may equally well lead to the apocalyptic self-destruction of the species. This he counters by saying that, while others may be guided by the inevitable decay implicit in natural processes, he is concerned with the regenerative and the emergent.

Reflective questions: Leonardo da Vinci

Throughout this journey into collective thinking, our prototype synthesist has been Leonardo da Vinci. In his work that remains publicly available, we find him invariably drawing on all the knowledge cultures and answering the full set of commonsense questions. In his journal, his commitment to reproducing the whole never wavers. He was rarely satisfied with his own attempts to achieve his vision, and abandoned them if he considered them unsatisfactory (or not having life, as Alexander would say). He often lost interest in a machine, a drawing or an idea once he had captured the essence of it for himself. His supposed self-portrait in figure 10.5 shows a person of fierce integrity.

Leonardo's life story is a set of paradoxes. He was born to an unmarried peasant woman, and therefore fell outside the social framework of the time. His father took him into his well-to-do household, legitimating his existence through a custom of the day. As an apprentice he was recognised from the first as far beyond his master.

Figure 10.5 Leonardo da Vinci *Self-portrait.* **Turin, Royal Library.**

In another era this might have earned him envy and ostracism. But in the climate of the Renaissance it brought immediate recognition and his own studio. He was the first since Grecian times to represent the sacred in everyday human form, thereby challenging the rule of church and state. Yet rather than being burnt as a heretic, the quality of the works brought him powerful friends from both camps. He never compromised in his work itself, but he was not above engaging in politics, and arguing his own case. He was a social isolate and the object of some ferocious envy, even from Michelangelo. Yet he was always surrounded by close friends and given enormous respect by the general populace.

Leonardo's life and work embodied the advice from today's Renaissance thinkers. His introspective writing in his journal allowed him to develop his individual perspective on the relationship between his work and the physical and social worlds in which he lived. His passion for empirical observation gave great realism to his paintings, and success to his technological inventions. His theoretical reflections on the physical world foreshadowed fractals and chaos theory. His reflections on human beings in relation to their social setting documented the drive towards 'material complexification' (the external world of machines) and 'psychic interiorisation' (the interior world of personal reflection) identified by Theilhard de Chardin as characterising the human condition.

Every aspect of his life crossed the divisions that, as Foucault observes, make up the power structures of every society. He used his own moment of grace to the fullest, basing his ethical principles on his responsibility to faithfully represent the whole spirit of each synthesis he undertook. Aesthetically, he was only satisfied when he had fully captured the pattern that connects. This is the driver that Bateson argues is biologically pre-determined in us all. The adequacy of his capacity for reflectivity, for learning from his own learning in all dimensions, is confirmed in the awed admiration for his work today, half a millennium later. Figure 10.6 is but one of many examples of Leonardo using both sides of his brain at once. The journal notes and the birds in flight illustrate his innate drive to combine the empirical, social, ethical,

Figure 10.6 Leonardo da Vinci *Manuscript of the flights of birds*. **Turin, Royal Library.**

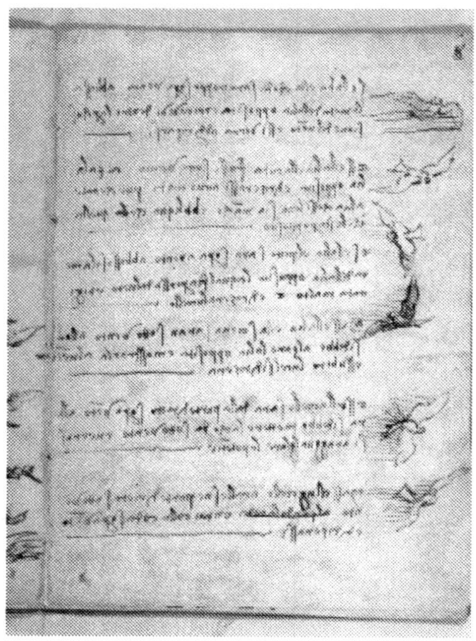

aesthetic, introspective and reflective as one collective thought.

All of this does not mean that life was easy for Leonardo da Vinci. His childhood on the edge of his society, the envy of some of his peers, the risks from pushing at the boundaries of what could then be known, all took their toll. At various stages he had no patrons, some of his work was destroyed or abandoned by others, and he felt isolated from his society. His journals reveal that he felt that he had only a few close friends that he could trust, and he never established a standard family relationship. Many of these feelings of being separate from their society, from the outside looking in, are also found in the autobiographies of others who took the road of synthesis.

If Leonardo had not existed, this century would have had to invent him. In his own times and ever since that time, he acts as a symbol of the capacity of all humans to capture the spirit of their time, and to change the direction of their times. His very existence stands as a record for those wishing to follow the same vision today.

Implications

The conclusions from our search for the skills of synthesis that underlie collective thinking are clear. Anyone seeking to guide social change will find a most powerful ally in a synthesis of the decision-making knowledge cultures, and the combined answers to common-sense questions. It has been possible to put together a portfolio of tools for social change towards synthesis and sustainability, extending each of the knowledge cultures and optimising the human capacity for collective knowledge. Figures 10.7 and 10.8 provide diagrammatic summaries of collective thinking as an open learning process.

Figure 10.7 Essential content of the collective learning spiral (Brown 2007)

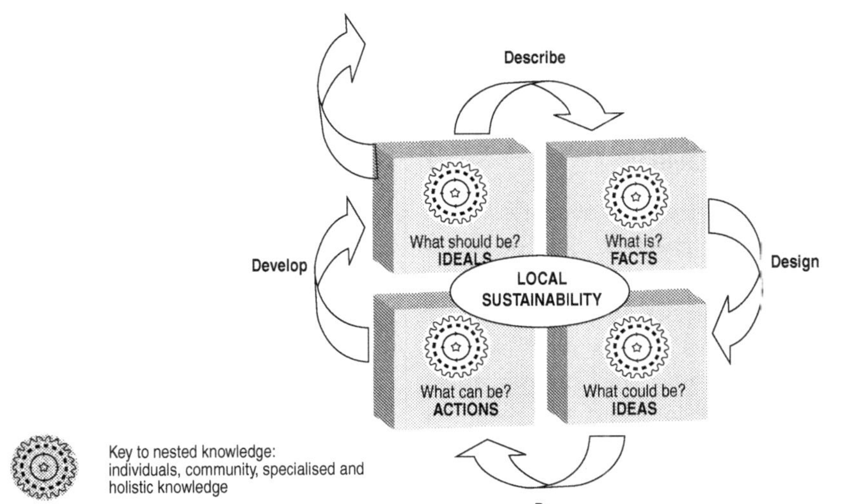

© Valerie A. Brown

Figure 10. 8 Essential process of the collective learning spiral (Brown. 2007)

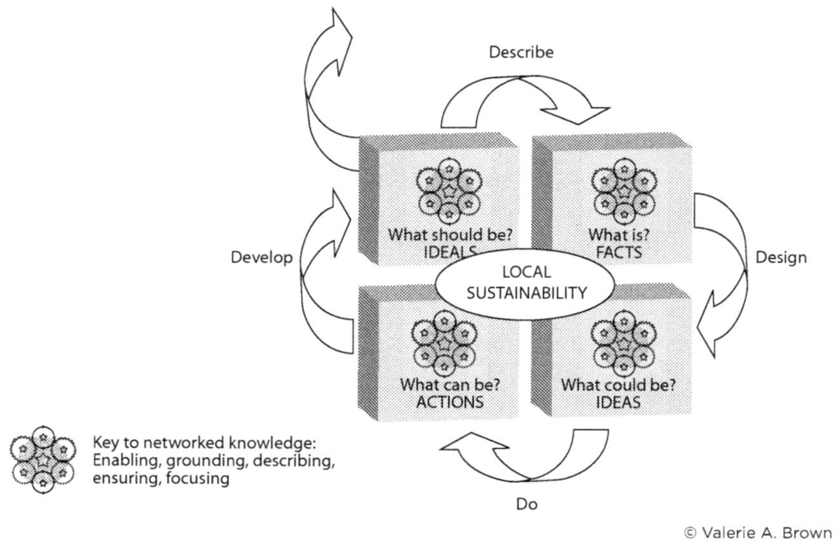

There are three options in applying those tools. First, those who pick them up can adopt synthesis as their preferred mode of knowledge and action within their own area of expertise. They have the rich battery of tools from which to draw, within the increasing repertoire of collective knowledge strategies. They will be warmly welcomed. There is increasing acceptance in every specialised field that such a contribution is badly needed.

Second, since they will almost certainly have been trained in one particular specialised skill, they can provide the specialist expertise that any synthesis will need. This means working in a team with others who do not speak your language nor share your means of inquiry. If they have adopted synthesis as their main goal, they will offer a synthesising framework and a common language. This requires the practitioner to accept that their role is partner to the other knowledge cultures, not competitor.

Or, third, they can take acting collectively, both personally and as an advocate for synthesis for the whole society, as their main goal. Since this is still an exploratory field, those who choose to go in that direction will find that they are also the pathfinders. They will be joining an increasing company of practitioners from all fields who are working with, rather than from within, the disparate knowledge cultures. Depending on where they are working, they may be labelled a utopian thinker, a radical activist, a valued change agent, a genius, or an impractical dreamer. They may be considered divisive, and disloyal to their profession, community or organisation. But, as with Leonardo, if they are convinced that they can best work with collective thinking and action, they will not give in.

We are witnessing the beginning of what may be a knowledge revolution, a move from fragmented to unified decision-making. This is a bold claim, but one supported by the evidence in the preceding chapters. In this last chapter we revisited some of the prophets of a sustainable humane future, and evaluated how far we have come towards their dream. The question remains, will this knowledge revolution lead us to a sustainable, human future? The answer is that, for the sake of the world, it must.

The last word can remain with a poet:

"Human eyes impose a human pattern,
Decipher constellations against
featureless dark.
All's fire, said Heraclitus; measures of it
kindle as others fade. All changes,
yet all's one.
We are born of the ethereal fire
and we return there.
Understand the Logos; reconcile
opposing principles.
Round earth's circumference and
atmosphere
Bombs and warheads crouch
waiting their time.
Well, Greek, we have not found the
road to virtue.
I shiver by the fire this winter's day.
The play of opposites, their
interpenetration –
there's the reality, the fission and the
fusion.

Judith Wright, 1989, from *A Human Pattern: Selected Poems*. Angus and Robertson Modern Poets Sydney pp241-242.

Bibliography

Abercrombie MLJ 1967 *The anatomy of judgment: an investigation into the processes of perception and reasoning.* Harmondsworth, Penguin. p15

Alexander, C. 2002a. *The nature of order. Book one: the phenomenon of life*. Berkeley, The Centre for Environmental Structure. p106

Alexander, C. 2002b. *The nature of order. Book two: the process of creating life.* Berkeley, The Centre for Environmental Structure.

Alexander C, 2003. *The nature of order. Book three: a vision of the living world.* Berkeley, The Centre for Environmental Structure.

Alexander, C. 2005. *The nature of order. Book four: the luminous ground.* Berkeley, The Centre for Environmental Structure.

Alexander, C., Ishikawa, S., Silverstein, M. Jacobson, M., Fiksdahl-King, I. and Angel, S. 1977 *A pattern language*. New York: Oxford University Press.

Andrews, N. and Tyson, L. 2004 The upwardly global MBA. *strategy+business magazine*, Fall, 2004.

Aronsen, H. 2004 *Buddhist practice on western ground: reconciling eastern ideals and western psychology*. Shambhala Publications, Boston, USA.

Aslin, H.J. and Brown, V.A. 2002 *Terms of engagement: community engagement in the Murray-Darling Basin*. Research Report. Canberra, Bureau of Rural Sciences.

Aslin H.J. and Brown V.A. 2004 *Towards whole-of-community engagement: a practical toolkit*. Canberra, Murray-Darling Basin Commission.

AtKisson, A. 1999 *Believing Cassandra: an optimist looks at a pessimist's world*. White River Junction, VT, Chelsea Green Publishing Company.

Barbarazi, A-L 2002 *Linked: The new science of networks.* Cambridge, Mass., Perseus.

Baran, P. quoted in Buchan, M.

Barthes, R. 1972 *Mythologies*. New York, Hill and Wang.

Bateson, G. 1972 *Steps Toward an Ecology of Mind.* St Albans: Paladin, 1973, 57-58

Bateson, G. and Bateson, M.C. 1987 *Angels Fear: towards an epistemology of the sacred.* New York: Macmillan Publishing.

Baton Forum An activist group of local sustainability practitioners, founded 2001. Founding executive: Victoria Critchley, Gary Hankinson, Judy Lambert, Jenny Scott, Gary Smith (chair), Greg Walkerden. Collaborative researchers, Sonia Duus and Julia Pickforth. Sydney, Institute of Environmental Studies, UNSW.

Bawden, R. 1995 I as in academy: learning to be systemic. *Systems Research journal* 12(3): 229-238.

Bawden, R. 1998 *Systemic development: a learning approach to change*. Hawkesbury, NSW, Department of Systemic Development, University of Western Sydney.

BBC News Online 2000 *Live Aid: The show that rocked the world.* (from BBC News Online April 5, 2000). BBC News Online.

Beck, U. 1992 (1986) *The risk society: towards a new modernity.* Trans. Mark Ritter. London, Sage Publications.

Belenky, M, Clinchy B., Goldberger, N., and Tarule, J. 1997. *Women's ways of knowing* Basic Books, New York

Berger, P. L. and T. Luckmann. 1971 *The social construction of reality: a treatise in the sociology of knowledge*, Harmondsworth, Penguin.

Berlin, I. 1990 *The crooked timber of humanity. Chapters in the history of ideas*, edited by H. Hardy. New York, Random House

Berlin, I. 1997. The hedgehog and the fox in *Against the current: essays in the history of ideas*, London, Pimlico, p12.

Berry, M., J. Garrard and D. Greene. 1994 *Reducing lead exposure in Australia. Volume 1: An assessment of impacts. Final report. Volume 2: technical appendices risk assessment and analysis of economic, social and environmental impacts*,

Canberra, National Health and Medical Research Council.

Berry, T. 1999 *The great work: our way into the future*. New York, Bell Tower.

Blackburn, S. 1994 *The Oxford dictionary of philosophy*. Oxford, Oxford University Press.

Blake, W. 1978 Ages of innocence. In AP Wavell ed. *Other men's flowers* London, Cape p15

Bohm, D. 1996 *On dialogue*. London, Routledge

Bohr, N. 1955 *The unity of knowledge*. New York, Doubleday & Co.

Brian, B. 1995 *The voice of genius. Conversations with Nobel scientists and other luminaries*. Cambridge, Mass., Perseus.

Briggs, J. and Peat, D. 1998 *Seven life lessons of chaos: spiritual wisdom from the science of change*, New York, Harper Collins Publishers.

Brown, D. 1998 *The da Vinci code* Random House, New York.

Brown, L. 2005 *Vital signs 2005* Washington, Worldwatch Institute.

Brown, V.A. 1981 From sickness to health: an altered focus for health-care research. *Social Science and Medicine* 15A, 195-201

Brown, V.A. 1985 Brown, V.A. Towards an epidemiology of health: A basis for planning community health programs. *Health Policy and Education* (Elsevier), 331-340

Brown, V.A. 1984 In Formation: The place of specialists, generalists and holists in an information service. In, Proceedings, Melbourne Eighth General Assembly and Congress of the International Federation for Documentation Commission for Asia and Oceania (FID/CAO).

Brown, V.A. 1997. *Managing for local sustainability: policy, problem-solving, practice and place*. Canberra, National Office of Local Government

Brown, V.A. 2001a Planners and the planet: reshaping the people/planet relationship: do planners have a role? *Australian Planner* 38 3 67- 73.

Brown, V.A 2001b Monitoring changing environments in Environmental Health. *Environmental Health*, 1.1, p21-34

Brown V.A. 2003 The symbols of my place are the foundations of my world: towards a shared Australian sense of place In Cameron,J. (ed) *Changing Places: Re-imagining Australia*. Double Bay NSW, Longueville books.

Brown, V.A. 2005 Knowing: linking the knowledge cultures of sustainability and health. In Brown V.A., Grootjans, J., Ritchie, J. , Townsend, M. and Verrinder, G. (eds) *Sustainability and Health: supporting global ecological integrity in public health*. Sydney, Allen and Unwin. and London, Earthscan.

Brown V.A. 2006 Towards the next Renaissance? Collective decision-making combining community, specialised and organisational knowledge. *International*

Journal of Knowledge, Culture and Change Management Volume 6, Issue 3. 43-56

Brown, V.A. 2007 *Leonardo's vision. A guide to collective thinking and action*. Rotterdam, SENSE.

Brown V.A., Griffith R. and Ohlin J. 1998 *A fine place to live with just a few things to fix* in D. Broom ed. Health Development Papers. Canberra, National Centre for Epidemiology and Health.

Brown, V.A., Grootjans, J., Ritchie, J., Townsend, M. and Verrinder, G. 2005. *Sustainability and health: supporting global ecological integrity in public health. Sydney,* Allen and Unwin and London, Earthscan.

Brown, V.A. and Pitcher, J. 2005 Linking community and government: Islands and beaches. In

Keen, M, Brown, V.A. and Dyball, R. eds.,*Social learning in environmental management*, London, Earthscan. 123-146.

Brundtland, G. 1986 *Our common future.* Commission for the Future, Melbourne, World Commission on the Environment and Development, Oxford University Press.

Buchan, M. 2002 *Small worlds and the ground-breaking science of networks.* New York, W. W. Norton.

Buber, M. 2002 (1947) *Between man and man.* London, Routledge Classics

Buber, M. and Smith,R. G. 1999 (1937) *I and Thou.* Edinburgh, T. & T. Clark

Burke, E. (Womersley, D. ed) 1999 (1757) *A philosophical enquiry into the sublime & beautiful & other pre-revolutionary writings.* Penguin, UK.

Buzan, T. 1988 *Make the most of your mind*, New York, Simon & Schuster. (Orig. 1977)

Calder, N. 1970. *Technopolis: social control of the uses of science*. New York, Simon and Schuster.

Cameron, J. (ed) *Changing Places: Re-imagining Australia.* Double Bay NSW, Longueville books.

Campbell, J. 2001 *The liar's tale: a history of falsehood.* New York, Norton

Camus, A. 1937 *Betwixt and between*. Trans P. Cody, London, Hamish Hamilton p162

Capra, F. 1984 *The turning point: science, society, and the rising culture.* New York, Bantam Books.

Capra, F. 1996 *The web of life: A new synthesis of mind and matter.* London, Harper Collins

CERES. 2004. Our Work: The CERES Principles. CERES. http://www.ceres.org/our_work/principles.htm (accessed 7/12/2004)

Checkland, P. 1981. *Systems thinking, systems practice*, Chichester, John Wiley & Sons.

Churchman, C.W. 1979 *The systems approach and its enemies*. New York: Basic Books

Clark, K. 1935 Introduction. In *Leonardo da Vinci.* London Penguin.

Clark, R.W. 1973 *Einstein. The life and times.* London, Hodder and Stoughton.

Commission for Sustainable Development 1997. Agenda 21: Work Program 1993-98. Thematic Clusters, New York, United Nations.

Computer Professionals for the Public Sphere 2006 http://www.cpsr.org/ [accessed September 2006]

Cool Communities Program material.

Available at http://www.google.com.au/ - [2/04/05]

Available at http://www.google.com.au/ - [2/04/05]

Available at: http://www.unep.org/geo2000/ov-e/index.htm [8/5/005)

Available at: www.greenhouse.gov.au/coolcommunities (8/5/05)

Costanza, R. and Jorgensen, S.E. Eds. 2002. *Understanding and solving environmental problems in the 21st century: Toward a new, integrated hard problem science.* Kidlington, Elsevier.

Da Vinci - see Leonardo da Vinci

De Chardin - see Teilhard de Chardin

De Montaigne, M. 1990 (1575) *Collected essays.* London, Penguin Classics

Dante, A. 1982 *The Divine Comedy 3: Paradiso.* (Tr. J. Sinclair). Oxford, Oxford University Press.

Dening, G. 1980 *Islands and beaches: discourse on a silent land. A case study of the Marquesas 1774-1880* Honolulu, University Press of Hawaii.

Descartes, R. 1946 (1637) *A discourse on the method of rightly conducting one's reason.* (Tr. J. Veitch) London, Everyman's Library 570.

Development Gateway 2001a. *Editorial Policy URL* http://www.developmentgateway.org/EditorialPolicy [5/1/02]

Diamond, J. 1991 *The rise and fall of the third chimpanzee*. London, Vintage.

Diamond, J, 1999 *Guns, germs and steel: the fates of human societies.* New York, WW Norton.

Dickens, C. 1992 (1854) *Hard times.* New York, Knopf and Random House.

Dower, N. 2003. *An Introduction to global citizenship.* Edinburgh, Edinburgh University Press

Duchamp, M. 1961 *Apropos of readymade,* [http: post-dogmatism-arts.net/museum/ collage/readymades.html] accessed 2004 Source: The international museum of collage, assemblage and construction.

Duchamp, Marcel (1887-1968) replica 1950 (1917) *Fountain.*
Philadelphia, Philadelphia Museum of Art. 2005. © Photo SCALA Florence Code A193803

Dwyer, B. 2002 *Aspects of governance and public participation in the remediation of the Murray-Darling Basin.* Unpublished Ph D thesis. Richmond NSW, University of Western Sydney.

Dyball, R., Beavis, S., and Kaufman, S. 2005 Complex adaptive systems: models of social learning and sustainability, in M.Keen, V.A. Brown, and R. Dyball eds. *Social Learning in Environmental Management.*, London, Earthscan.

Eames, R. 2005. Partnerships in civil society: linking, bridging and bonding social capital. In

Keen. M, Brown, V and Dyball, R. eds. *Social Learning in Environmental Management.* Earthscan, London. 78-91

Earth Charter initiative 2003 *The Earth Charter principles.* The Earth Charter initiative. http://www.earthcharter.org/earthcharter/ charter.htm [2/2/03]

Environmental Health Commission (UK). 1997 *Agendas for Change.* Chartered Institute of Environmental Health, Chadwick House Group, London. 7-15

Falk, R. 1999 *Predatory globalization : a critique.* Cambridge, Polity Press.

Fisher, F. 2002 Scared of GM? An environmental scientist responds to "GM scare hots up", Letter to the Editor. Melbourne, Sunday

Fernandez-Armesto, F. 2001 *Truth: a history and a guide for the perplexed.* New York, St. Martin's Griffin.

Flannery, T. 1994 *The Future Eaters: An ecological history of Australasian lands and people,* Chatswood, Sydney, Reed Books.

Flecker, James Elroy 1947 (1900) *Selected Poems.* London, The Richards Press. p 45.

Foucault, M. 1970 *The order of things: an archaeology of the human sciences,* New York, Pantheon Books.

Foucault, M. 1979 *Power, Truth, Strategy.* (Morris and Patton, eds) Melbourne, Prometheus Books. = contains excerpts from *Discipline and punish* and *The clinic.*

Friere, P. 1996 (1970) *Pedagogy of the oppressed.* London, Penguin.

Frost, R. 1972. *Selected poems.* Harmondsworth, Penguin. p46.

Gang, P. and Morgan, M. 2004 Rules of Dialogue. In *TIES Curriculum documents,* Mass. Endicott College.

Gardner, H. 1983 *Frames of Mind: the theory of multiple intelligences.* New York, Basic Books.

Gardner, H. 1993 *Multiple intelligences. The theory in practice.* New York, Basic books.

Geertz, C. 1973. *The interpretation of cultures- selected essays.* New York, Basic Books.

Geertz, C. 1993 (1983). *Local Knowledge.* London, Fontana. p89

Geldof, Bob - see BBC News Online

Google Search Engine http://www.google.com.au/ [30.2.07]

Gould, S. J. 2003. *The hedgehog, the fox, and the magister's pox: mending the gap between science and the humanities.* New York, Harmony Books. p260

Greene, D. 2001 *Where have all the policies gone?: analysis of policy implementation in Australia using lead as a case study.* New York, Poola Foundation.

Greene, D., M. Berry, and J. Garrard. 1993 *Reducing lead exposure in Australia: Risk assessment and analysis of economic, social and environmental impacts*. Canberra: National Health and Medical Research Council.

Griffith, R. 2002 *How shall we live? The sustainability agenda and institutional change in local governments in Australia.* Unpublished PhD thesis. Richmond NSW, University of Western Sydney.

Grootjans, J, Brown VA, and Ritchie, J. and Verrinder,G. 2002 Sustainable development: A cornerstone of public health. *Environmental Health* 2. 1. 71-75

Gunderson, L. H. and Holling, C. S. 2002 *Panarchy: Understanding transformations in human and natural systems*. Island Press, Washington.

Habermas, J. 1971 Knowledge and human interests. (Tr. J. J. Shapiro) Boston. Beacon Press.

Hajer, M. 1996 Ecological modernisation as cultural politics in S. Lash, B. Szerszynski, and B. Wynne eds *Risk, Environment and Modernity*. London, Sage. 246-269

Hancock, T. 1986 Vision workshop design. Healthy Cities project, Copenhagen, WHO Europe

Heisenberg, W. http://www-gap.dcs.st-and.ac.uk/~history/Mathematicians/Heisenberg.html [accessed 20.5.06]

Hesse, H. 2002 (1943) *The glass bead game:(Magister Ludi)*. (Tr. R. T. Winston and C. T. Winston) New York, Picador.

Hospers. J. 1969 *Aesthetics*. New York, The Free Press.

Hughes, R. 1980 *The shock of the new: art and the century of change*. London: British Broadcasting Corporation (8 part television series).

Ife, J. 1995 *Community development: creating community alternatives – vision, analysis and practice*, Melbourne, Longman.

Illich, I. 1973 *Deschooling society*, Harmondsworth: Penguin

Illich, Ivan. 1973 *Tools for conviviality*. New York, Harper & Row

Jacobs, J. 1972 (1961) *The death and life of great American cities*, Harmondsworth, Penguin.

Jacobs, J. 2000. *The Nature of Economies,* Toronto, Random House

Jung, C. 1964 *Man and his symbols*. Doubleday, New York.

Keen, M, Brown, V.A. and Dyball, R. 2005 *Social learning in environmental management*, London, Earthscan.

Kelly, G. A. 1985 (1965) *Theory of personality: The psychology of personal constructs*, New York. W.W. Norton and Company.

Kelman, D. 1975 Dimensions of social change. New York, Routledge.

Knowles, M. S. 1980 *The modern practice of adult education: from pedagogy to andragogy,* Englewood Cliffs: Prentice Hall/Cambridge.

Koestler, A. 1971. *Koestler on creativity*, Videorecording, New York, Time-Life Multimedia.

Koestler, A. 1990 *The act of creation*, London, Penguin.

Kolb, D.A. 1984 *Experiential Learning: Experience as the Source of Learning and Development*, Englewood Cliffs, Prentice-Hall.

Kolb, D.A., Lublin, S. and Spoth, J. 1986 Strategic management development: using experiential learning theory to asses and develop managerial competencies" *Journal of Management Development* (5) 3: 13-24.

Kolb, D. A., Rubin, I. M. & McIntyre, J.M. 1974 *Organizational psychology: an experiential approach*, Englewood Cliffs, Prentice Hall.

Kuhn, T. 1970 *The Structure of Scientific Revolutions*. Chicago, University of Chicago Press.

Latour, B and Woolgar, S. 1986 *Laboratory Life: The Construction of Scientific Facts*. Princeton, Princeton University Press.

Lee, L. 1959 *Cider with Rosie*. London, The Hogarth Press. p245

Leonardo da Vinci. 1998 (orig. 1498). *The Notebooks of Leonardo da Vinci.* I Richter ed. Oxford, Oxford University Press. pp 6, 14, 38, 281, ...

Figure: Leonardo da Vinci (1452-1519) *Drawings of figures, probably studies for the Adoration of the Magi* (facsimile) Florence, Uffizi. © 1990 Photo SCALA, Florence. Code 0062432

Figure 1.2 Leonardo da Vinci (1452-1519) *Mona Lisa.* Paris, Louvre. © 1990 Photo SCALA, Florence. Code: 0080332

Figure 2.1. Leonardo da Vinci (1452-1519) August. 1473 *Tuscan Landscape* Florence, Uffizi. © 1990 Photo SCALA, Florence. Code: 0091464

Figure 3.3 Leonardo da Vinci (1452-1519) *Five grotesque heads*, inv. 227. Venice, Accademia. © 1900 Photo SCALA, Florence. Code: 0064602

Figure 5.3 Leonardo da Vinci (1452-1519) *Head of man with scheme of proportions* Venice, Academia. © 1990 Photo SCALA, Florence. Code: 0064600

Figure 6.1 Leonardo da Vinci (1452-1519) *Plan of the City of Imola* from the Codex Windsor (facsimile). Vinci, Museo Vinciano. © 1990 Photo SCALA, Florence. Code: 0067413

Figure 7.1 Leonardo da Vinci (1452-1519) *Study of the varying effects of light striking a profile from a single source* (facsimile) Florence, Uffizi. © 1990 Photo SCALA. Florence. Code: 0062453

Figure 7.4 Leonardo da Vinci (1452-1519) c1472-1519 *Sketch of a perpetual motion device*. London, Ann Ronan Picture Library. © 2005 Photo SCALA, Florence. Code h500989

Figure 9.1 Leonardo da Vinci (1452-1519) *Annunciation*. Florence, Uffizi. © Photo SCALA, Florence. Code: 0118677

Figure 10.5 Leonardo da Vinci (1452-1519) *Self-Portrait no. 15741.* Turin, Biblioteca Reale (Royal Library). © 1990 Photo SCALA, Florence. Code: 0044263

Figure 10.6 Leonardo da Vinci (1452-1519) *Manuscript on the flight of birds*, f. 8 recto. Turin, Biblioteca Reale (Royal Library). © 1990 Photo SCALA, Florence. Code: 0044160

Leopold, A. 1989 (1949) Thinking like a mountain. In *A sand country almanac* New York, Oxford University Press. p132

LETSLINK UK. 2004 LETSLINK UK: UK LETS and Complementary Currencies Development Agency. LETSLINK, UK

Local Sustainability Project, 1992 - . Collaborative action research program, V. A. Brown, Director. Centre for Resource and Environmental Studies , Australian National University 1992-1996; Regional Integrated Monitoring Centre University of Western Sydney 1996 - 2002; and School of Resources, Environment and Society, ANU, 2002 -

Lorenz E. 1972 *Predictability: does the flap of a butterfly's wing in Brazil set off a tornado in Texas?* Paper delivered at the Washington DC, 1972 conference of the American Association for the Advancement of Science.

Lovelock, J. 1979 *Gaia: A New Look at Life on Earth.* Oxford, Oxford University Press.

Lovelock, J. 2000 *Homage to Gaia. The life of an independent scientist*. Oxford, Oxford University Press.

Luria, A. R., 2004 *The man with a shattered world: the history of a brain wound.* (Tr. L. Solotaroff) Boston, Harvard University Press.

Macchiavelli, N. 1976 (1512) *The Prince.* (Tr. G. Bull), Harmondsworth, Penguin Books (Penguin classics).

Mandelbrot, B. 1983 (Orig. 1977) *The Fractal Geometry of Nature.* New York, W.H. Freeman and Company.

McMichael, A. J. 2001 *Human frontiers, environments and disease. Past patterns, uncertain futures.* Cambridge, Cambridge University Press.

McMichael. A.J. 1999 From hazard to habitat: rethinking environment and health. *Epidemiology* 10. 461-464

Mead, M. 1978 *Culture and commitment : the new relationships between the generations in the 1970s.* New York, Columbia University Press.

Meadows, D. H., Meadows, D. L., and Randers, J. 1992 *Beyond the Limits: confronting global collapse, envisioning a sustainable future,* Vermont, Chelsea Green. pp ix, 5

Meadows, D. H., Meadows, D. L., Randers, J. and Behrens WW. 1972 *The Limits to Growth* New York, Universe.

Medawar, P. B. 1967 (1915) *The art of the soluble.* London, Methuen.

Metzner, R. 1999 *Green psychology - transforming our relationship to the Earth.* Vermont, Park Street Press.

Butor, M. 1960 *Passing time.* New York, Jupiter Books. p69

Midgely, G. 2000 *Systemic intervention: philosophy, methodology and practice.* New York, Kluer.

Milgram, S. *1992 The Individual in a Social World: Essays and Experiments.* (J. Sabini and M. Silver, eds) New York, McGraw-Hill.

Mill, J. S. 1995 (1859) *On liberty.* Bartleby.com.

Millennial Ecological Assessment. 2004 *Biodiversity and Human Well-being: A Synthesis Report for the Convention on Biological Diversity.* www.MAweb.org [30.2.07

Montessori, M. 2007 (1912) *The Montessori Method* (Tr Anne Everett George) University of Pennsylvania digital library. www.digital.library.upenn.edu/women/montessori/method/method.html [30.2.07].

More, T. 1995 (1530). *Utopia* (Tr. G.M. Logan, R.M. Adams, C. Miller) Cambridge, Cambridge University Press. www.d-holliday.com/tmore/utopia.htm [2.10.05]

Morgan, D. 1967. Must art tell the truth? *Journal of Aesthetics and Art Criticism* 26 Fall Issue pp 17-27

Mumford, L. 1961. *The City in History. Its Origins, its Transformations and its Prospects.* New York, Harcourt, Brace & World.

Nanno, J. 2003. Engineering, biology, integrity, thanksgiving and original instructions: Modes of reasoning in the management of Canada/US Haudenosaunee. Conference paper for *Global Ecological Integrity, Human Rights, and Human Responsibilities: Intersections Between International Law and Public Health,* Urbino, Italy, June 26-July 1, 2003.

National Health and Medical Research Committee NHMRC 1995 *Environmental Health Committee reports. 1989-95.* NHMRC, Federal Department of Health, Canberra

National Research Council (U.S.) 1999 *Our Common Journey: the transition toward sustainability.* Washington, DC, Board on Sustainable Development, Policy Division, National Academy Press.

Nicholson, R., Stephenson, P., Brown, V.A. and Mitchell, K. eds. 2002 *Common ground and common sense: community-based environmental health action.* Part A. Making a difference. Sydney, Local Sustainability Project, University of Western Sydney.

Nix, H. 1995 Water Issues in Australia. Keynote address to National Conference on Water Issues in Australia. Canberra, Centre for Resource and Environmental Studies, Australian National University.

Nowotny, H., Gibbons, M., Limoges, C., Schwartzman, S., Scott, P., and Trow, M. 1994 *The new production of knowledge.*

The dynamics of science and research in contemporary societies. London: SAGE.

O'Toole. J. 1993 *The executive's compass: business and the good society*. London, Oxford University Press.

Owen, H. 1997 *Open space technology: a user's guide*. San Francisco, Berrett-Kohler.

OECD 2003 *OECD Environmental Indicators: development, measurement and use*. Paris, Organisation for Economic Co-operation and Development.

OED Oxford English Dictionary. 1995. Oxford, Oxford University Press

Open Knowledge Network, http://dgroups.org/groups/OKN/docs/intellectual_property_rights.doc [1/10/05]

Orwell, G. 1970 (Orig.1958) *Collected essays, journalism and letters*. Eds Sonia Orwell and Ian Angus, London, Penguin.

Paine, G. 2005 Patterns, wholes and sustainable development: closing the gap between concern and action. Unpublished PhD thesis, Richmond NSW, University of Western Sydney.

Paine, T. 2004 (1776) *Common Sense. Of the origin and design of government in general*. Available from http://www.ushistory.org/paine/commonsense/ [3/12/2005].

Parissi, C. 2003 Lessons from local sustainability monitoring: outcomes and aftermath of the Western Sydney Regional State of the Environment Report 2000. *Environmental Health* Vol 3, No 3, 54-66.

Pattern Languages for Living Communication, a program of Computer Professionals for the Public Sphere. See http://www.cpsr.org/program/sphere/patterns/ [15/10/05].

Pattern Languages for Public Comunication Group (http://www.pattern-language.lists.cpsr.org) is a working group of Computer Professionals for the Public Sphere (http://www.cpsr.org).

Piaget, J. 1951 *The Child's Conception of the World.* London, Routledge, Kegan Paul.

Picasso, Pablo (1881-1973) Three Musicians, (Fontainebleau, estate 1921). New York, Museum of Modern Art (MoMA). © 2006 Photo SCALA Code: 0121112.

Plato 1991 (360 B.C) *The republic of Plato.* ed. A. Bloom. New York, Basic Books.

Plato, on Socrates 1962 (360 BC)*The Republic.* (Tr. RG Bury) Cambridge, Mass, Harvard University Press

Polanyi, M. 1958 *Personal Knowledge*. Chicago, The University of Chicago Press.

Port Pirie case study: Field notes from author's site visit, November 2003.

Prigogine, I. and Stengers, I. 1984 *Order out of Chaos*, London, Fontana.

Putnam, R. 1995 Bowling alone: America's declining social capital. *Journal of Democracy vol. 6, no. 1, 65-78.*

Putnam, R. 2000 *Bowling Alone: The Collapse and Revival of American Community. New York,* Simon and Schuster.

Ravetz, J. 1999 Citizen Participation for Integrated Assessment: new pathways in complex systems. Int. *Journal of Environment and Pollution* 11.3:331-350; special issue on citizen participation.

Ravetz, J. R. 1999 What is Post-Normal Science, *Futures*, 31(7):647-653

Rees, S. and Rodley, G. 2004 *The human costs of managerialism*. Melbourne, Pluto Press Australia.

Rees, W and Wackernagel M. 1996 *Our ecological footprint : reducing human impact on the earth*, Gabriola Island, B.C, New Society Publishers.

Rees, W. 1992 Ecological Footprints and appropriated carrying capacity. What urban economics leaves out. *Environment and Urbanisation*, 4, 3, 121-130.

Reid, J. 2002 *Factor of 10. Art exhibition*. Canberra, School of Art, Australian National University.

Rittel, H. and Webber, M. 1973 Dilemmas in a General Theory of Planning. *Policy Sciences* 4: 155-169

Robert Frost. 1972 Mending Wall *Selected Poems* Harmondsorth, Penguin p 43.

Rosetta Stone: The Rosetta Stone is a stone carved in 196BC, found in a small village in the Nile Delta called Rosetta (Rashid) with writing on it in two languages (Egyptian and Greek), using three scripts (hieroglyphic, demotic and Greek). The three scripts allowed the priests, government officials and rulers of Egypt to read what it said

Rykiel, E.J. Jr., Berkson, J.M., Brown, V., Krewitt, W., Peters, I., Schwartz, M., Shogren, J. and van der Molen, D. 2002 Science and decision-making. In R. Costanza and S.E. Jorgensen [Eds.] *Understanding and solving environmental problems in the 21st century: Toward a new, Integrated "hard problem science."* Kidlington, England, Elsevier. pp 153-166

Sahani M. 2003. Separate realities: community-based environmental management for health from the inside out and the outside in. Unpublished PhD thesis. Richmond NSW, University of Western Sydney.

Santini, Pier. 1972 *Modern landscape painting* London, Phaidon Press

Saul, J. Ralston. 1992 *Voltaire's bastards: the dictatorship of reason in the West*. New York, Free Press.

Schirmer, J. 2000. Plantation forestry disputes: Case studies on concerns, causes, processes and resolution, *CRC-SPF Technical Report No. 42*, Canberra, Cooperative Research Centre for Sustainable Production Forestry.

Schuler, D. 2001 Cultivating society's civic intelligence: patterns for a new 'world brain', *Information, Communication and Society* 4, 2, Summer issue.

Schuler, D. 2002 A pattern language for living communication: a global participatory project. In PDC '02 Participatory Design Conference proceedings. Palo Alto, California, CPSR.

Schuler, D. 2007 *Liberating voices. A pattern language for civic communication*. Cambridge, Mass.,The MIT Press.

Scott B., Szerszynski, S., and Wynne, B. eds. 1996 *Risk, Environment and Modernity. Towards a New Ecology.* London, Sage.

Scott, P. 1984 *The Raj Quartet: The Jewel in the Crown; The Day of the Scorpion; The Towers of Silence; A Division of the Spoils.* London, William Morrow.

Senge, P. 1994 *The fifth discipline: the art and practice of the learning organization*. New York, Doubleday/Currency.

Shiva, V. 1989 *Staying Alive. Women, Ecology, Development*. London, Zed Books.

Singer, P. 2002. *One World: the ethics of globalisation*, Melbourne, Text.

Smithson, M. 1989. The changing nature of ignorance, and Managing in an age of ignorance, In *New Perspectives on Uncertainty and Risk*, J. Handmer (ed) Canberra, Centre for Resource and Environmental Studies, ANU p.5-66

Smuts, J. C. 1936 *Holism and evolution* (Third edition). London, Macmillan.

Smuts, J.C. 1931. *London Times 7/7*, 2.9.31

Stafford, K. 1999 Places and stories In Metzner, R. 1999. *Green psychology- transforming our relationship to the Earth.* Vermont, Park Street Press.

Sun Tzu, (ed. Wing, R. L). 1988 (600 BC) *The art of strategy: a new translation of Sun Tzu's classic, The art of war*. New York, Doubleday.

Teilhard de Chardin, T. 1975 (1955) *Le Phenomene Humain (The Phenomenon of Man).* (Tr. B. Wall) New York, Harper & Row.

The Economist print edition. 2004 *The lunatic you work for* Economist print edition May 6, 2004. www.Economist.com.

Thoreau, H. D. 1960 (1854) *Walden and civil disobedience*. Boston, Houghton Mifflin.

Titmuss, R. 1976 (1950) *Problems of social policy.* London, Greenwood.

Toulmin, S. 1958 *The uses of argument.* Cambridge, Cambridge University Press.

Toulmin, S. 1972 *Human Understanding Vol 1.* Oxford, Clarendon.

Toulmin, S. 1977 From form to function: The philosophy and history of science in the 1950's and now. *Studies in Contemporary Scholarship. Daedelus, Summer*, 1. 106 3. 143-162.

Ulrich, W. 1983 *Critical heuristics of social planning: a new approach to practical philosophy.* Berne, Haupt Press.

UNCED (United nations conference on environment and development) 1992 *Agenda 21.* Geneva, United Nations http://www.un.org/esa/sustdev/documents/agenda21/index.htm. [30/4/2005].

United Nations conference on environment and development (UNCED) 1993 *Local Agenda 21: Chapter 28.* New York, Commission for Sustainable Development.

United Nations Environment Program. 2003 *Global Environment Outlook 2003*, New York, United Nations (Available on-line at http://www.unep.org/Geo/index.htm [18/09/04].

University of Nevada. 2004. Resources for the study of beowulf. Reno, University of Nevada, http://www.library.unr.edu/subjects/guides/beowulf.html#poetry [8/12/2004].

van der Velden, M. 2005 Programming for cognitive justice: Towards an ethical framework for democratic code. *Interacting with computers* 17, 1, 105-120

Vickers, G. ed. 1972 *A classification of systems. Yearbook of the Society for General Systems* Washington, DC, Research/Academy of Management Research.

Voltaire, F. 1998 (1759) Candide or optimism, in *Candide and other stories.* (Tr. R. Pearson) Oxford University Press, London.

Wackernagel, M. and Rees, W. 1996. *Our ecological footprint: reducing human impact on the earth*, Gabriola Island, BC ; Philadelphia,

Waring, M. 1999 (1988) *Counting for nothing: what men value and what women are worth.* Toronto, University of Toronto Press.

Wark, M. "Third Nature", *Cultural Studies*, 1994, 8:115-32

Watson, J. 1968. *The Double Helix: A personal account of the discovery of the structure of DNA*, New York, Scribner.

Wenger, E. 1998. *Communities of Practice: Learning, Meaning, and Identity* Cambridge University Press, Cambridge

Wilson, E. O. 1998. *Consilience: the unity of knowledge*. Knopf, New York

Witelson, S. F., Kigar, D.L., & Harvey, T. 1999 The exceptional brain of Albert Einstein. *Lancet*, 353, 2149-2153.

Whitman, W. 1996 (1855). *A choice of Walt Whitman's verse*, Ed Donald Hall, Faber pp27,28

World Bank. 1999. *World Development Report 1998/99: Knowledge for Development.* New York, Oxford University Press.

World Bank. 2002. *Sustainability Report*, World Bank, New York

World Commission on Dams. 2000. Dams and development: A new framework for decision-Making. The Report of the World Commission on Dams (released November 16 2000) http://www.dams.org/report/ (accessed 6/12/2004)

World Resources Institute, United Nations Environment Program, United Nations Development Program and the World Bank. 2000 . *World Resources 2000* Oxford University Press, New York

Wright J. 1996 (1976). Eve to her daughters. In *A Human Pattern. Selected Poems* Sydney Angus and Robertson p134

Wright J. 1996 (1976). Fission and fusion. In *A Human Pattern. Selected Poems* Sydney Angus and Robertson p

WSROC (Western Sydney Regional Organisation of Councils). *2001. Regional State of the Environment Report*. Sydney: Western Sydney Regional Organisation of Councils, Blacktown and University of Western Sydney. www.uws.edu.au/rimc/

Wynne, B.1996 May the Sheep Safely Graze? A Reflexive View of the Expert-lay Knowledge Divide. In Scott Lash, Bronishlaw Szerszynski and Brian Wynne eds. *Risk Environment and Modernity. Towards a New Ecology.* London: Sage Publications. York

Yes Minister Jay, A. and Lynn, J. 1980 BBC, UK, broadcast 1983 and 1984. http://www.bbc.co.uk/comedy/guide/articles/y/yesminister_7777145.shtml [20/09/04]

List of figures

List of boxes

Acknowledgments

Leonardo's Vision is a celebration of the mutual learning that comes with collective thinking. Everyone listed below is already putting Leonardo's vision of a collective mind into practice. My heartfelt thanks go to all those who have traveled with me along this road.

Thank you to:

Partners in teaching/learning relationships. I have learnt so much from fellow-learners in courses in ecology, human ecology, social ecology, adult learning, community education, public health and environmental health, in the universities of Queensland, Latrobe, New South Wales, Tribhuvan Nepal, Western Sydney and the Australian National University. In particular, the Human Ecology Forum at the Australian National University, and the on-line program in integrated studies from TIES (The Institute of Educational Studies), Endicott College, USA, justified my faith that there is such a thing as collective learning.

Loyal readers of draft after draft. My unending gratitude to those willing to read the early versions of *Leonardo's Vision*. In order of timing they have been Wallace Brown, Bina Brown, Erica Fisher, Jan Ritchie, Julia Pickforth, John Grootjans, Greg Walkerden, Glenda Verrinder, Greg Walkerden, Sally McArthy, John Harris, Wendy Rainbird, Brett Odgers, Meg Switzer and Karlson (Charlie) Hargrove. The insights they provided are now part of the book. My early editor, Trish Boekel, gave generously of her considerable professional skill in helping me learn to translate research reports into a book. My later editor, Cameron Cooper of Coopermedia, helped me take the book the last mile.

Fellow members of synoptic workshops. The book is enriched by the case studies of collective thinking developed through a series of synoptic workshops. I have gained enormously from working in the following teams:

- Individual knowledge: The PhD seminar group of the Local Sustainability Project, University of Western Sydney, which included Rod Griffith, Brian Dwyer, Greg Paine, Jenny Scott, Fred Lehmann, Patrick Poon and Mazrura Sahani, all whom developed their own synoptic research programs, and eventually told their own tales.
- Community knowledge: The team who worked on the manual *Common ground and commonsense,* Peter Stephenson, Rosemary Nicholson and Kathy Mitchell, were my valued colleagues at the University of Western Sydney for seven years. Peter and Kathy later translated the manual for indigenous use. The *Western Sydney Regional State-of-the-Environment Report* involved Jocelyn Powell, Marion Benjamin, Marnie Campbell, Adrienne Murphy, and Amanda Walmsley. The *Towards whole-of-community engagement* toolkit was a co-production with Heather Aslin and Alison Reid.
- Specialised knowledge. My fellow-editors of the public health text, *Sustainability and health,* John Grootjans, Jan Ritchie, Mardi Townsend, and Glenda Verrinder, gave me three years of the most rewarding and stimulating companionship. Colin Soskolne, Alan AtKisson, Lee Harpur and Tony McMichael were expert contributors to this endeavour. With fellow-editors of *Social Learning in environmental*

management, Meg Keen and Rob Dyball of the Australian National University, we managed to introduce synthesis into the heartland of monodisciplinarity.

- Organisational knowledge. Changing whole organisations is a notoriously difficult task. My congratulations to Victoria Critchley for her courage in developing a whole-of-council change strategy. From this emerged the Baton Forum of local government sustainability practitioners, whose social change program continues with flair and commitment. The original team was Garry Smith, Judy Lambert, Skye Addison, Sonia Duus, Jenny Scott, Gary Hankinson, Greg Walkerden and Victoria Critchley.
- Holistic knowledge. Committed holists on whose work I have drawn are the Human Ecologist Stephen Boyden founder of Human Sciences; performance artist John Reid; computer scientist Douglas Schuler, author of *Liberating Voices. A pattern language for living communication*; and urban planner Greg Paine, whose PhD thesis was on *Parts, wholes and sustainable development*.

Institutional support. This book is the outcome of a ten year project that could not have proceeded without grants from the Australian Research Council Linkage Program, the national Public Health Education and Research Program and the Rockefeller Foundation. Jackie Ohlin and Kathy Mitchell, managers of the Local Sustainability Project at the Centre for Resource and Environmental Studies, Australian National University, and Regional Integrated Monitoring Centre, University of Western Sydney, respectively, proved that organisational knowledge is both an art and a science. Peter Kanowski, Head of the School of Resources, Environment and Society, Australian National University, generously provided the institutional base where I completed the book. The Renaissance environment and the intellectual challenge of the multidisciplinary centre Villa Sirmione of the Rockefeller Foundation at Bellagio, where I was a scholar-in-residence in 1999, gave rise to the original ideas.

My family and the future. Children and grandchildren focus the mind powerfully on the need to have a future worth living. The arrival of our children, Matthew, Bina, and AJ changed me from a clinical research scientist to a passionate advocate for a collective responsibility for social change. For grandchildren Christopher, Sarah, Elliot and Amon, the change seems even more urgent. The unfailing personal support and life partnership of Robin Wallace Brown made the whole project possible.

My most sincere thanks to you all. Valerie A. Brown, Canberra, August 2006

Index

Leonardo's Vision: A guide to collective thinking and action is part of the

TRANSGRESSIONS: CULTURAL STUDIES AND EDUCATION Series
Volume 16

Scope

Cultural studies provides an analytical toolbox for both making sense of educational practice and extending the insights of educational professionals into their labors. In this context *Transgressions: Cultural Studies and Education* provides a collection of books in the domain that specify this assertion. Crafted for an audience of teachers, teacher educators, scholars and students of cultural studies and others interested in cultural studies and pedagogy, the series documents both the possibilities of and the controversies surrounding the intersection of cultural studies and education. The editors and the authors of this series do not assume that the interaction of cultural studies and education devalues other types of knowledge and analytical forms. Rather the intersection of these knowledge disciplines offers a rejuvenating, optimistic and positive perspective on education and educational institutions. Some might describe its contribution as democratic, emancipatory, and transformative. The editors and authors maintain that cultural studies helps free educators from sterile, monolithic analyses that have for too long undermined efforts to think of educational practices by providing other words, new languages, and fresh metaphors. Operating in an interdisciplinary cosmos, Transgressions: Cultural Studies and Education is dedicated to exploring the ways cultural studies enhances the study and practice of education. With this in mind the series focuses in a non-exclusive way on popular culture as well as other dimensions of cultural studies including social theory, social justice and positionality, cultural dimensions of technological innovation, new media and media literacy, new forms of oppression emerging in an electronic hyperreality, and postcolonial global concerns. With these concerns in mind cultural studies scholars often argue that the realm of popular culture is the most powerful educational force in contemporary culture. Indeed, in the twenty-first century this pedagogical dynamic is sweeping through the entire world. Educators, they believe, must understand these emerging realities in order to gain an important voice in the pedagogical conversation.

Without an understanding of cultural pedagogy's (education that takes place outside of formal schooling) role in the shaping of individual identity—youth identity in particular--the role educators play in the lives of their students will continue to fade. Why do so many of our students feel that life is incomprehensible and devoid of meaning? What does it mean, teachers wonder, when young people are unable to describe their moods, their affective affiliation to the society around them. Meanings provided young people by mainstream institutions often do little to help them deal with their affective complexity, their difficulty negotiating the rift between meaning and affect. School knowledge and educational expectations seem as anachronistic as a ditto machine, not that learning ways of rational thought and making sense of the world are unimportant.

But school knowledge and educational expectations often have little to offer students about making sense of the way they feel, the way their affective lives are shaped. In no way do we argue that analysis of the production of youth in an electronic mediated world demands some "touchy-feely" educational superficiality. What is needed in this context is a rigorous analysis of the interrelationship between pedagogy, popular culture, meaning making, and youth subjectivity. In an era marked by youth depression, violence, and suicide such insights become extremely important, even life saving. Pessimism about the future is the common sense of many contemporary youth with its concomitant feeling that no one can make a difference.

If affective production can be shaped to reflect these perspectives, then it can be reshaped to lay the groundwork for optimism, passionate commitment, and transformative educational and political activity. In these ways cultural studies adds a dimension to the work of education unfilled by any other sub-discipline. This is what Transgressions: Cultural Studies and Education seeks to produce—literature on these issues that makes a difference. It seeks to publish studies that help those who work with young people, those individuals involved in the disciplines that study children and youth, and young people themselves improve their lives in these bizarre times.

Volume 1
An Unordinary Death...*The Life of a Suicide Bomber*
Khalilah Christina Sabra
Paperback ISBN 90-77874-36-4 Hardback ISBN 90-77874-37-2

Volume 2
Lyrical Minded
The Critical Pedagogy of Hip-Hop Artist KRS-ONE
Priya Parmar, *City University of New York, USA*
Paperback ISBN 90-77874-50-X Hardback ISBN 90-77874-64-X

Volume 3
Pedagogy and Praxis in the Age of Empire
Towards a New Humanism
Peter McLaren & Nathalia Jaramillo
Paperback ISBN 90-77874-84-4 Hardback ISBN 90-77874-85-2

Volume 4
Dewey and Power
Renewing the Democratic Faith
Randall Hewitt, *University of Central Florida, USA*
Paperback ISBN 90-77874-92-5 Hardback ISBN 90-77874-93-3

Volume 5
Teaching, Learning and Other Miracles
Grace Feuerverger, *Ontario Institute for Studies in Education, University of Toronto, Canada*
Paperback ISBN 90-77874- Hardback ISBN 90-77874-

Volume 6
Soaring Beyond Boundaries
Women Breaking Educational Barriers in Traditional Societies
Reitumetse Obakeng Mabokela (ed.), *Michigan State University, USA*
Paperback ISBN 90-77874-97-6 Hardback ISBN 90-77874-98-4

Volume 7
Diasporic Ruptures
Globality, Migrancy, and Expressions of Identity
Volume I
Alireza Asgharzadeh, *York University, Canada,* Erica Lawson, *The University of Nevada, Las Vegas, USA,* Kayleen U. Oka, *University of Toronto, Canada and* Amar Wahab (Eds.), *University of Toronto, Canada*
Paperback ISBN 90-8790-050-3 Hardback ISBN 90-8790-051-1

Volume 8
Diasporic Ruptures
Globality, Migrancy, and Expressions of Identity
Volume II
Alireza Asgharzadeh, *York University, Canada,* Erica Lawson, *The University of Nevada, Las Vegas, USA,* Kayleen U. Oka, *University of Toronto, Canada and* Amar Wahab (Eds.), *University of Toronto, Canada*

Volume 9
The Politics of Education
An Introduction
Tony Monchinski, *City University of New York, USA*
Paperback ISBN 90-8790-037-6 Hardback ISBN 90-8790-038-4

...me 10
...age, Capital, Culture
...tudies of Language and Education in Singapore
...ish, S. Gopinathan and Yongbing Liu, *Centre for Research in Pedagogy and Practice, ...al Institute of Education, Nanyang Technological University, Singapore*
...back ISBN 978-90-8790-122-6 Hardback ISBN 978-90-8790-123-3

...ume 11
...e Great White North?
...xploring Whiteness, Privilege and Identity in Education
...aul Carr, *Youngstown State University, USA* and Darren Lund (Eds.), *University of Calgary, Canada.*
Paperback ISBN 978-90-8790-142-4 Hardback ISBN 978-90-8790-143-1

Volume 12
Education under Occupation
The Heavy Price of Living on a Neocolonized and Globalized World
Pierre Orelus, *University of Massachusetts-Amherst, USA.*
Paperback ISBN 978-90-8790-145-5 Hardback ISBN 978-90-8790-146-2

Volume 13
The Ethics of Caring
Bridging Pedagogy and Utopia
Tammy A. Shel
Paperback ISBN 978-90-8790-209-4 Hardback ISBN 978-90-8790-210-0

Volume 14
Disrupting Privilege, Identity, and Meaning
A Reflective Dance of Environmental Education
Alison Neilson
Paperback ISBN 978-90-8790-182-0 Hardback ISBN 978-90-8790-183-7

Volume 15
Girls in a Goldfish Bowl
Moral Regulation, Ritual and the Use of Power amongst Inner City Girls
Rosalyn George, *Goldsmiths, University of London, UK*
Paperback ISBN 978-90-8790-185-1 Hardback ISBN 978-90-8790-186-8